DUDLEY OBSERVATORY AT ALBANY, N.Y.

POPULAR ASTRONOMY.

A

CONCISE ELEMENTARY TREATISE

ON THE

SUN, PLANETS, SATELLITES AND COMETS.

BY

O. M. MITCHEL, LL.D.,

DIRECTOR OF THE CINCINNATI AND DUDLEY OBSERVATORIES.

PHINNEY, BLAKEMAN & MASON,

No. 61 WALKER STREET.

1860.

Entered, according to Act of Congress, in the year 1860, by

O. M. MITCHEL,

In the Clerk's Office of the District Court for the Southern District of New York.

STEREOTYPED BY
SMITH & McDOUGAL,
82 & 84 Beekman-st., N. Y.

PRINTED BY
J. D. BEDFORD & CO.,
115 & 117 Franklin St.

PREFACE.

The author has no other apology to present for offering to the public the following work on "Popular Astronomy" than the marked favor with which his "Planetary and Stellar Worlds" has been received, both in this country and in Europe.

The science of Astronomy is so rapidly progressive that to keep the public advised of its advances new works are required almost every year. This may be offered as an additional reason for the present publication.

In the preparation of the work I have availed myself of so many sources of information that it would be quite impossible for me to specify the authors or the volumes to which I am indebted. The plan and the cast is all my own. I have endeavored to follow the path of real discovery, and in every instance to present the facts and phenomena so as to afford to the reader and student an opportunity to exercise his own genius in their discussion and resolution, before offering the explanation reached by ancient or modern science. It is hoped that this method of treating the subject which is new, (so far as I know,) may avail in exciting a greater interest in the examination of those great problems of the universe whose successful solution constitutes the chief honor of human genius.

In a few instances I have ventured to present the results of my own observations, and have occupied a short space in exhibiting a sketch of new methods and new instruments, which have been introduced into the observatories at Cincinnati and at Albany.

DUDLEY OBSERVATORY, *January*, 1860.

CONTENTS.

CHAPTER I.

THE SUN, THE CENTRAL ORB OF THE PLANETARY SYSTEM.

CHAPTER II.

MERCURY, THE FIRST PLANET IN THE ORDER OF DISTANCE FROM THE SUN.

CHAPTER III.

VENUS, THE SECOND PLANET IN THE ORDER OF DISTANCE FROM THE SUN.

CHAPTER IV.

THE EARTH AND ITS SATTELLITE; THE THIRD PLANET IN THE ORDER OF DISTANCE FROM THE SUN.

CHAPTER V.

MARS, THE FOURTH PLANET IN THE ORDER OF DISTANCE FROM THE SUN.

CHAPTER VI.

THE ASTEROIDS: A GROUP OF SMALL PLANETS, THE FIFTH IN THE ORDER OF DISTANCE FROM THE SUN.

CHAPTER VII.

JUPITER, ATTENDED BY FOUR MOONS, THE SIXTH PLANET IN THE ORDER OF DISTANCE FROM THE SUN.

CHAPTER VIII.

SATURN, THE SEVENTH PLANET IN THE ORDER OF DISTANCE FROM THE SUN, SURROUNDED BY CONCENTRIC RINGS, AND ATTENDED BY EIGHT SATELLITES.

CHAPTER IX.

THE LAWS OF MOTION AND GRAVITATION.

CHAPTER X.

THE LAWS OF MOTION AND GRAVITATION APPLIED TO A SYSTEM OF THREE REVOLVING BODIES.

CHAPTER XI.

INSTRUMENTAL ASTRONOMY.

CHAPTER XII.

URANUS, THE EIGHTH PLANET IN THE ORDER OF DISTANCE FROM THE SUN.

CHAPTER XIII.

NEPTUNE, THE NINTH AND LAST KNOWN PLANET IN THE ORDER OF DISTANCE FROM THE SUN.

CHAPTER XIV.

THE COMETS.

CHAPTER XV.

THE SUN AND PLANETS AS PONDERABLE BODIES.

CHAPTER XVI.

THE NEBULAR HYPOTHESIS.

INTRODUCTION.

THE great dome of the heavens, filled with a countless multitude of stars, is beyond a doubt the most amazing spectacle revealed by the sense of sight. It has excited the admiration and curiosity of mankind in all ages of the world. The study of the stars is therefore coeval with our race, and hence we find many discoveries in the heavens of whose origin neither history nor tradition can give any account. The science of Astronomy, embracing, as it does, all the phenomena of the celestial orbs, has furnished in all ages the grandest problems for the exercise of human genius. In the primitive ages its advances were slow, but by patient watching, and by diligent and faithful records transmitted to posterity from generation to generation, the mysteries which fill the heavens were one by one mastered, until at length, in our own age, there remains no phenomenon of motion unexplained, while the distances, magnitudes, masses, reciprocal influences, and physical constitution of the celestial orbs have been approximately revealed. In a former volume an attempt was made to trace the career of discovery among the stars, and to exhibit the successive steps by which the genius of man finally reached the solution of the great problem of the universe.

The performance of that task did not permit the special study of any one object, except so far as it was required in the march of the general investigation. It is our object now to execute what was then promised, and to examine in detail the various bodies which are allied to the sun, constituting (as we shall find) a delicately organized system of revolving worlds, a complex mechanical structure, whose stability has challenged the admiration of all thinking minds, and whose organization has furnished the most profound themes of human investigation.

The plan adopted will lead us to present clearly all the facts and phenomena resulting from observation; with these facts the student may exercise his own genius in attempting to account for the phenomena, before proceeding to accept the explanation laid down in the text.

To aid the memory and to present a systematic investigation, we shall adopt the simple *order* of *distance* from the solar orb, commencing with that grand central luminary, and proceeding outward from planet to planet, until we shall develop all the phenomena employed in the discovery of the great law of universal gravitation. With a knowledge of this law the worlds already examined cease to be isolated, and arrange themselves under the empire of gravitation into a complex system, the delicate relations of whose parts, leads to new discovery and to the final perfection of the system of solar satellites.

Having closed our investigation of the planets and their tributary worlds, we shall render an account of those anomalous bodies called comets, which, by the suddenness of their appearance, their rapid and eccentric motions, and the brilliant trains of light which sometimes attend them, have excited universal interest, not unattended with alarm in all ages of the world.

Before passing to the execution of this plan, we must examine, to some extent, the phenomena of the nocturnal heavens, as the stars furnish the *fixed* points to which all moving bodies are referred.

To the eye the heavens rise as a mighty dome, a vast hollow hemisphere, on whose internal surface the glittering stars remain forever fixed. In case we watch through an entire night, we find the groupings of stars slowly rising from the east, gradually reaching their culmination, and then gently sinking in the west. A more attentive examination enables the eye to detect some of these groups of stars toward the north which ever remain visible, rising, culminating, and descending, but never sinking below the horizon. Every star in this *diurnal revolution*, as it is called, is found to describe a *circle*, precisely as if the concave heavens were a hollow sphere to which the stars were attached, and that this hollow globe were made to revolve about a fixed axis, passing through its center. Indeed, we find by attentively watching, that this hypothesis of a spherical heavens accounts for all the phenomena already presented. As the stars are situated nearer to the extremity of the axis of revolution the circles they describe grow smaller and smaller, until, finally, we find *one* star which remains *fixed*, and this one must be at the point where the axis of the heavens pierces the celestial sphere. This is called the *north star ;* and the point in which the axis pierces the heavens is called the *north pole.* The opposite point is called the *south pole.*

Only one half of the celestial sphere is visible at one time above the horizon, but this spherical surface extends beneath the horizon, and forms a complete sphere, encompassing us on all sides, while its center seems to be occupied by the earth. It is true that, in the day-time, the stars fade from the sight in the solar blaze, but they are not lost; they still fill the heavens, as we shall see hereafter, and the starry sphere sweeps unbroken entirely round the earth.

These great truths, the diurnal revolution of the heavens, its spherical form, the central position of the earth, the north polar star, the axis of the heavens, the circles described by the stars, were among the discoveries of primitive antiquity, and are matters of the most simple observation.

The spherical form of the heavens was soon imitated, and the *artificial* globe became one of the first astronomical instruments. On this artificial globe certain lines were drawn to imitate those described in the heavens by the celestial orbs, and as these lines must henceforth form a part of our language we proceed to give the following *Definitions*:—

A *great* circle is one whose plane passes through the center of the sphere.

A *small* circle is one whose plane does not pass through the center of the sphere.

The *axis* of the heavens is an imaginary line passing through the center of the earth, and about which the heavens appear to revolve once in twenty-four hours.

A *meridian* is a great circle passing through the highest point of the celestial sphere (called the *zenith*) and the axis of the heavens.

The *equator* or *equinoctial* is a great circle, perpendicular to the axis of the heavens, and half-way between the north and south polar points.

These important lines have been employed from the earliest ages in the study of the heavenly bodies, and having thoroughly mastered their meaning and position we are prepared to examine any changes of location which may be discovered among the vast multitude of shining bodies which go to fill up the concave of the celestial sphere.

We shall proceed, then, without further delay, to the execution of the plan already laid down.

CHAPTER I.

THE SUN, THE CENTRAL ORB OF THE PLANETARY SYSTEM.

DISCOVERIES OF THE ANCIENTS.—THE SOURCE OF LIFE AND LIGHT AND HEAT.—THE SUN'S MOTION AMONG THE STARS.—HIS ORBIT CIRCULAR.—LENGTH OF THE YEAR.—INEQUALITY OF THE SUN'S MOTION.—EXPLAINED BY HIPPARCHUS.—SOLAR ECLIPSES.—THEIR FIRST PREDICTION.
DISCOVERIES OF THE MODERNS.—THE SUN'S DISTANCE.—HIS HORIZONTAL PARALLAX.—IMPORTANCE OF THIS ELEMENT.—MEASURED BY THE TRANSIT OF VENUS.—THE SUN'S ACTUAL DIAMETER AND REAL MAGNITUDE.—HIS ROTATION.—THE SOLAR SPOTS.—THEIR PERIODICITY.—SPECULATIONS AS TO THE PHYSICAL CONSTITUTION OF THE SUN.

THE sun is beyond comparison the grandest of all the celestial orbs, of which we have any positive knowledge. The inexhaustible source of the heat which warms and vivifies the earth, and the origin of a perpetual flood of light, which, flying with incredible velocity in all directions, illumines the planets and their satellites, lights up the eccentric comets, and penetrates even to the region of the fixed stars; it is not surprising that in the early ages of the world, this mighty orb should have been regarded as the visible emblem of the Omnipotent, and as such should have received divine honors.

On the approach of the sun to the horizon in the early dawn, his coming is announced by the gray eastern twilight, before whose gradual increase the brightest stars and even the planets fade and disappear. The coming splendor grows and expands, rising higher and yet higher, until, as the first beam of sunlight darts on the world, not a star or planet remains visible in the whole heavens, and even the moon, under this flood of sunlight, shines only as a faint silver cloud.

This magnificent spectacle of the *sunrise*, together with the equally imposing scenes which sometimes accompany the *setting* sun, must have excited the curiosity of the very first inhabitants of the earth. This curiosity led to a more careful examination of the phenomena attending the rising and setting sun, when it was discovered that the point at which this great orb made his appearance was not *fixed*, but was slowly shifting on the horizon, the change being easily detected by the observation of a few days. Hence was discovered, in the primitive ages—

THE SUN'S APPARENT MOTION.—In case the sun is observed attentively from month to month it will be found that the point of sunrise on the horizon moves slowly, for a certain length of time, toward the *south*. While this motion continues, the sun, at noon, when culminating on the meridian, reaches each day a point less elevated above the horizon, and the *diurnal arc* or daily path described by the sun grows shorter and shorter. At length a limit is reached; the point of sunrise ceases to advance toward the south, remaining stationary a day or two, and then slowly commences its return toward the north. This northern movement continues; each day the sun mounts higher at his meridian passage, the diurnal arc *above* the horizon grows longer and longer, until, again, a northern limit is reached, beyond which the sun never passes. Here he becomes stationary for one or two days, and then commences his return toward the south. Thus does the sun appear to vibrate backward and forward between his southern and northern limits, marking to man a period of the highest interest, for within its limits the *spring*, the *summer*, the *autumn*, and the *winter*, have run their cycles, and by their union have wrought out the changes of the *year*.

The length of this important period was, doubtless, first determined by counting the days which elapsed from the time when the sun rose behind some well-defined natural object in the horizon until his return in the same direction to the same point of rising. Of course, these changes in the sun's place were studied with profound attention. They were among the first celestial phenomena discovered, and among the first demanding explanation. The *stars* were found *never to change* their points of rising, culmination, and setting. Their diurnal arc remained forever the same, and the amount of time they remained above the horizon depended on their distance from the north polar point.

Observation having thus revealed the fact that the sun was undoubtedly moving alternately north and south, a more critical research showed the equally important truth, that this great luminary was slowly shifting its place among the fixed stars. This was not so readily determined; but by noting the brilliant stars which first appeared in the evening twilight, after sunset, it was soon discovered that these stars did not long remain visible. Indeed, the whole starry heavens seemed, from night to night, to be plunging downward to overtake the setting sun, or rather, that the sun himself was mounting upward to meet the stars, and thus was discovered a solar motion in a direction opposed to the diurnal revolution of the heavens.

From month to month the sun was seen to advance among the stars, and at the end of an entire year, after all the former changes of northern and southern motion had been accomplished, the sun was found to return to the same group of fixed stars from whence he set out; and thus it became manifest that this revolution among the stars was identical in period with the changes from north to south, and hence these phenomena had, in all probability, a common origin.

Here was the first great problem offered for solution to the old astronomers. The facts and phenomena were carefully studied, and the reader may now exercise his own power of thought in an effort to explain the facts recorded, before accepting the solution we are about to present.

An examination of the points of rising, of culmination, and of setting, of the *fixed* stars, showed them to be absolutely invariable, and in case these glittering points could leave behind them, in their diurnal revolution, lines of silver light, sweeping upward from their point of rising to the meridian, and downward to the point of setting, these lines of light would be seen to be *parallel circles*. All the stars *north* of the equinoctial (in the region of the earth *we* inhabit) describe diurnal circles, of which more than one half is *above* the horizon, while all the stars *south* of the same line sweep round in circles, of which *less than half* lies above the same plane. Any star, precisely on the equinoctial, half way between the north and south poles, passes one half its revolution *above* and the other half *below* the horizon.

These facts being carefully noted, it was seen that in case the sun, on any day of its annual journey, chanced to coincide with a fixed star, that for that day the sun and star would describe the same diurnal circle, and would remain above the horizon an equal length of time. Thus along the sun's path it became possible to select a number of stars over which the sun passed, and which would by their position mark his route in the heavens. To aid in this investigation, as well as for some other purposes, the ancients erected a *vertical staff* on a level plane, and then noted where the shadow of the top of the staff fell at noon each day throughout the year. This instrument was called *a gnomon,* and its use revealed many important facts in the solar motion, and detected others hitherto overlooked. If on the same day we note carefully the length of the shadow of the gnomon a little while before and after noon, we shall find the shadow slowly decrease in length as the sun rises to its culmination; and immediately after passing the meridian the shadow commences to increase in length. Mark the point where the *shortest shadow fell,* and the line joining this point with the foot of the gnomon is a *north and south* line, and on this line all the *noon* shadows will fall throughout the entire year.

By a careful examination it was discovered that the *noon shadow* on the day of the *winter solstice* or southern limit always fell on the same point. The same was true of the noon shadow on the day of the *summer solstice* or northern limit. These points were exactly opposite each other on the sun's *apparent orbit* (or *path among the stars*). It was further discovered, that selecting any day in the year, the noon shadow for that day invariably fell on the same point as it had done on preceding years; and hence it became manifest that the sun's track among the stars *did not change* from year to year.

The question now arose as to the figure of the sun's path: was that figure a *circle?* and did the sun move with *uniform* velocity? As all the stars described diurnal circles; as this curve was the simplest as well as the most beautiful of curves; as its curvature was every where the same; as it had neither beginning nor ending, it was early adopted as the *celestial curve*, shadowing forth, even in its form, the ceaseless journeys of the revolving worlds. It was assumed then that the sun swept round the sphere of the heavens once a year, with uniform velocity, in a circular orbit, of which the earth was in the center.

This hypothesis accounted fully for all the discovered phenomena, and justly ranks among the most important of the primitive discoveries.

The *gnomon* gave to the old astronomers a ready means not only of tracing the sun's path among the stars, but also of measuring the inclination of the plane of the *ecliptic* to the plane of the *equinoctial*. This is readily seen from the subjoined figure.

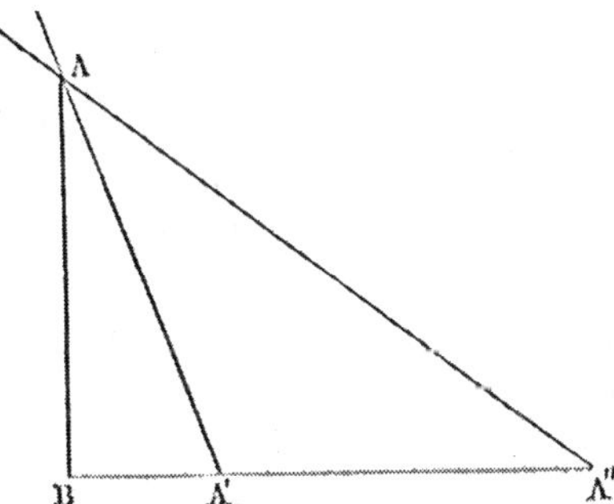

Let AB represent the gnomon, A′ the shadow of the vertex at noon on the day of the summer solstice, and A″ the shadow at noon on the day of the winter solstice. Then will the angle A′AA″ measure the entire motion of the sun from north to south; and as one half of this motion lies north and the other half south of the equinoctial, it follows that half the angle, A′AA,″ measures the inclination of the ecliptic to the equinoctial.

In the earliest ages it was assumed that the sun's orbit was absolutely fixed among the stars, and that the points in which this circle crossed the equinoctial were in like manner invariable. These points of intersection are of the highest importance. That one through which the sun passes in going from south to north, is called the *Vernal Equinox*, while the opposite point, through which the solar orb passes in going from north to south, is called the *Autumnal Equinox.* On the day of the equinoxes, as the sun's center was then on the equinoctial, the diurnal arc described by the sun would lie one half above, and the other half below the horizon, making the length of day and night *precisely equal.*

Among the ancient nations the day of the vernal equinox was an object of especial interest, as it heralded the coming of spring, and its approach was marked by the rising of a certain bright star in the early dawn of the morning. Now, in case the vernal and autumnal equinoxes were invariable, the same star by its *heliacal* rising, (as it was called,) would mark the crossing of the equinoctial by the sun in the spring and the equality of day and night. After the lapse of few centuries it was discovered, by the length of the noon shadow of the gnomon, that the sun had reached the equinoctial point, and yet the sentinel star did not make its appearance. Either the equinox or the star was in motion. It was soon decided that the vernal and autumnal equinoxes are both slowly moving backwards along the equinoctial, and thus the sun crosses this celestial circle each year a little behind the point of the preceding year.

The ancient nations all seem to have attained to a

knowledge of this great truth, and some of them are said to have fixed the period in which the vernal equinox retrogrades around the entire heavens, a period of nearly twenty-six thousand years; as this is a matter of simple observation, and as the rate of motion can be obtained by comparing recorded observations, made at intervals of four hundred or five hundred years, we may readily credit the statement that this period became known even anterior to the commencement of authentic history.

This discovery of the retrocession of the equinoxes led to a more critical examination of the sun's apparent motion. This motion had been assumed to be *uniform*, and in case this hypothesis could be maintained, the solar orb ought to occupy an equal amount of time in passing over the two portions of its orbit north and south of the equinoctial, that is, the number of days from the vernal to the autumnal equinox ought to be precisely equal to the number of days from the autumnal to the vernal equinox.

The Greek astronomer Hipparchus was the first to discover the important truth that an inequality existed in these two periods. He found from his own observations that the sun occupied *eight* days more in tracing the northern than it did in traversing the southern portion of its orbit. This was a discovery of the highest importance, as it seemed to involve the then incredible fact, that the lord of the celestial sphere, the great source of life, and light, and heat, traveled among the stars with a *variable* velocity.

In case the solar orbit was indeed a *circle*, this inequality of motion seemed to be impossible. The circular figure of the orbit could not be abandoned, neither was it possible on philosophical principles to give up the hypothesis of uniform motion. Here then was presented a problem of the deepest interest, to *preserve* the *circular* figure of the solar orbit and the *uniform motion* of the sun, and at the same time render a satisfactory account of the inequality discovered in the periods during which the sun remained north and south of the equinoctial. This problem was solved by Hipparchus; and before proceeding to examine the reasoning of the old Greek, let the student exercise his own genius in an attempt to explain the ascertained facts.

Hitherto it had been assumed, not only that the sun's orbit was circular, and that his motion was uniform, but also that the earth occupied the *exact center* of the circle in which the sun traveled round the heavens. By profound study Hipparchus discovered that all the facts could be explained by giving to the earth a position not in the center of the sun's orbit, but somewhat nearer to that portion of the solar orbit where his motion was most rapid. This will become evident from the figure. Let the circle A B C D represent the sun's circular orbit, in which the sun is supposed to move uniformly. This motion will only *appear* uniform to a spectator at the center O. If the observer be removed to O′, and the line E E′ be drawn perpendicular to O O′, the portion E′ A B E of the orbit will require a longer time for its description than the portion E C D E′, and hence in the former the sun will appear to move slower than in the latter. Indeed, it is manifest that the point V, on the line O O′ prolonged, is the place of swiftest motion, while the opposite point V′ is that in which the sun will appear to move slowest.

Hipparchus, not satisfied with thus rendering a general explanation of the phenomenon, undertook to determine the actual place of the earth inside the solar orbit, or the value of the distance O O′, which is called the *eccentricity.* Here is another problem for the examination of the student. It may be solved by simply knowing how

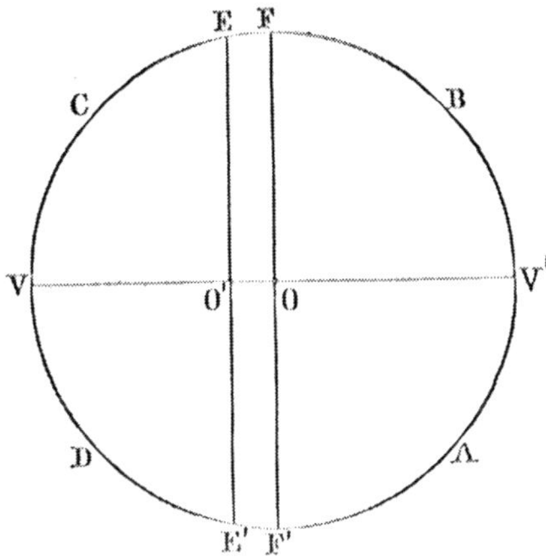

many days longer the sun remained north of the equinoctial than it did on the south of this circle. This quantity we have already given. By dividing the circle A B C D into as many equal parts as there are days in the year, and by drawing F F′ through the center O, and perpendicular to V V′, we have only to lay off from F to E half the excess in days, and draw E E′ parallel to F F′, and it will give at O′ the true place of the earth, and O O′ will be the eccentricity. An observer at O′ will see all phenomena actually detected in the sun's motion, while the circular orbit and uniform velocity are rigorously retained.

Having determined the earth's eccentricity, it was now

very easy to calculate the sun's place from day to day during his entire revolution among the fixed stars. This was actually done by the old astronomers; and as the computed places agreed with those observed within the limits of observation, with the rude instruments then in use, no further advance would be made in the solar motions.

ECLIPSES OF THE SUN.—No one has ever beheld the total disappearance of the sun in the day-time without a feeling of awe creeping through his frame, and, even now, when modern science predicts the coming of these amazing phenomena with unerring precision, a total eclipse of the sun never fails to inspire a certain feeling of gloomy apprehension. What, then, must have been the effect in the rude ages of the world of the fading out of the sun in mid-course through the heavens? Human genius, of course, bent all its energies to the resolution of the great problems involved in the occurrence of an eclipse of the sun. The first effort was directed to the discovery of the *cause* of these startling phenomena; and, this once determined, the second great effort was put forth to so master all the circumstances as not only to explain the eclipse but to *predict its coming.*

CAUSE OF A SOLAR ECLIPSE.—In searching for the cause by which the sun might be hidden, it was at once evident that there was but one object in the heavens sufficiently large to hide the whole surface of the sun. This body was the *moon.* Thus attention was directed to the lunar orb, and it was soon noticed that, while the bright stars and planets became visible in the darkness attending an eclipse of the sun, yet the brightest object in the heavens after the sun, was never visible during an eclipse. The moon was found to move among the stars with a velocity far greater than that of the sun. It was, moreover, seen that the moon's path crossed that of the sun twice during every revolution of the moon, and examining still more closely, it was discovered that no eclipse of the sun ever occurred except at the new moon. Now this rapidly revolving globe was evidently the nearest to the earth of all the heavenly bodies. It was seen, when a silver crescent, sometimes to pass over and hide the larger stars which fell in its path; it was also found that the moon, though invisible during a solar eclipse, always appeared immediately after very near the sun and as a slender crescent of light. These facts all combined to prove beyond a question that the sun was eclipsed by being *covered by the dark body of the moon.* The *cause* of the eclipse was thus reached, and it now remained to rob the phenomenon of its terrors by predicting when it might be expected.

To predict a solar eclipse with precision is a problem of great difficulty, even with the present extended knowledge of the laws and structure of the solar system. And yet we are informed that the old Greek astronomers succeeded in the resolution of this complex problem. This may have been done by long and persevering care in the record of these phenomena; for in case all the eclipses visible at any given place are recorded year after year for a period of *nineteen* years, it will be found that for the next period of nineteen years eclipses will happen on the *same days* and in the same order; so that an astronomer, whose diligence had been rewarded by the discovery of this grand truth, might acquire the highest renown among his countrymen and throughout the world by his superior wisdom in predicting the coming of an eclipse, though no special genius was put forth in the resolution of this great problem.

We are not quite certain, however, that the prediction of the first announced solar eclipse may not have been accomplished by the application of powerful thought and persevering observation. In case the effort were now made to predict a solar eclipse; as a starting point we know that no eclipse of the sun ever occurred except at the *new moon.* But at the time of a total eclipse of the sun the moon is interposed precisely between the eye of the observer and the sun, and a line joining the centers of these two great luminaries, produced to the earth, passes through the place of the observer. Hence, on the day and at the hour of an eclipse the *new moon* must be in the act of *passing from one side* of the sun's path to the other. To render an eclipse possible two conditions must be fulfilled at the same time; the moon must be *new,* and the moon's center must be in the act of *crossing the sun's orbit.* If the sun's annual route in the heavens were marked among the stars by a line of golden light, and the moon's motion be attentively watched, it will be found that at every one of her revolutions she crosses this golden line *twice.* The point of her crossing from south to north is called the moon's *ascending node,* while the point of crossing from north to south is the *descending node.*

These nodes do not remain fixed, but are in comparatively rapid motion, and finally accomplish an entire revolution around the heavens, *on the ecliptic.* If, then, we unite all these facts it will be seen that to produce to any observer an eclipse of the sun, the moon, at the *new,* must be exactly in one of her nodes, so that the center of the moon, the node, and the center of the sun, form one and the same straight line. Here, then, are the conditions precedent to a solar eclipse. It now remains to so follow these revolving orbs as to be able to anticipate the certain occurrence of these determined conditions. We follow, then, from night to night, the waning moon; she slowly approaches the sun; her light becomes a delicate crescent, just visible in the gray twilight of morning before the rising of the sun; at length the moon becomes invisible, and when she reappears it is on the opposite side of the sun, and her silver crescent of light is just above the setting sun. There was no eclipse because *this* new moon did not fall on the sun's path. It is, however, easy to mark the time of new moon, and equally easy to see and note the time when the moon is in her node, or on the ecliptic, and by thus watching, from new moon to new moon, we may see whether the interval from the passage of the node up to new moon is growing shorter, and at what rate it decreases, till, finally, we shall perceive that on the coming of a certain new moon *it* must fall precisely at the *node*, and on the day of this computed conjunction, to him who has watched, and waited, and pondered, and computed, the sun must fade away in total eclipse. Such is the train of reasoning and observation which may have first led to the resolution of this great problem, but to whose genius we are indebted for this grand discovery neither history nor tradition furnish any information.

In consequence of the near equality in the apparent diameters of the sun and moon, and a slight change in both due to a change of the actual distance from the earth (as will be shown hereafter), it sometimes happens that the moon's diameter is *less* than that of the sun. When this obtains during a solar eclipse there remains around the black disk of the moon a brilliant ring of solar light, and the eclipse is said to be *annular*. Whenever the moon's center, at the new, is not precisely at the node, but not so remote from it as the sum of the semi-diameters of those two orbs, there will be a *partial* obscuration of the sun.

We have presented these facts in this place, as known to the early astronomers, and as admirable means of exercising the power of thought on the part of those who may desire to devote themselves to the real study of the great phenomena of nature. We will recur to this subject again when we shall have mastered the laws of motion and of gravitation.

Such is a rapid survey of the discoveries of the ancients in the study of that great orb, which, from its splendor, even if it be a mere phantom of light, justly commands our admiration and deserves our best efforts to master its mysterious movements and its sublime phenomena.

We now proceed to exhibit those discoveries which could only be accomplished after man had armed himself with instruments of great power and delicacy, and with a vision increased a thousand fold beyond that, with which he is endowed by nature.

DISCOVERIES OF THE MODERNS.—The rude instruments employed by the early astronomers sufficed to fix the places of the sun and the other heavenly bodies with sufficient accuracy to give a general outline of the curves they described, and as these curves, as determined by observation, approximated the circular form, it was concluded that the deviations from that exact figure were only errors of observation. Knowing the period in which the sun revolves round the heavens, and the distance of the observer from the center of his assumed circular orbit, it was easy to compute accurately the sun's place among the stars on any day of the year. This computation being made, no instrument then in use could detect any difference between the *computed* place and that *actually held* by the sun. It was, therefore, unphilosophical to doubt the absolute truth of an hypothesis thus sustained by the best observations which could then be made. It was not at all difficult to observe roughly mere position, and any error of observation in fixing the place of the sun would, in the long run, be eliminated in its effects by taking into account a large number of revolutions. The degree of accuracy required in thus fixing the sun's place among the stars was widely different from that demanded in the

MEASUREMENT OF THE SUN'S DISTANCE.—The principles involved in the solution of this great problem were well understood by the old Greek astronomers, and were applied by them successfully in measuring the distances of inaccessible objects on the surface of the earth. These principles are so simple that a knowledge of the very first rudiments of geometry will suffice to render intelligible the methods which are employed in obtaining the data for computing the distances of the heavenly bodies.

Suppose it were required to learn the distance of the object A from the point C. From C send to A the visual ray C A, then lay off any line from C B perpendicular to A C, and measure its length. From B draw the visual ray B C, and measure the angle C B A. We have thus formed a right-angled triangle, in which the angle at C is a right angle, the *base*, C B, is known by measurement, and the angle C B A is known in the same way, hence may be computed, by the simplest elements of trigonometry, the length of the distance C A, or the required quantity.

Any error committed in the measurement of the angle C B A grows more powerful in its effect on C A, in pro-

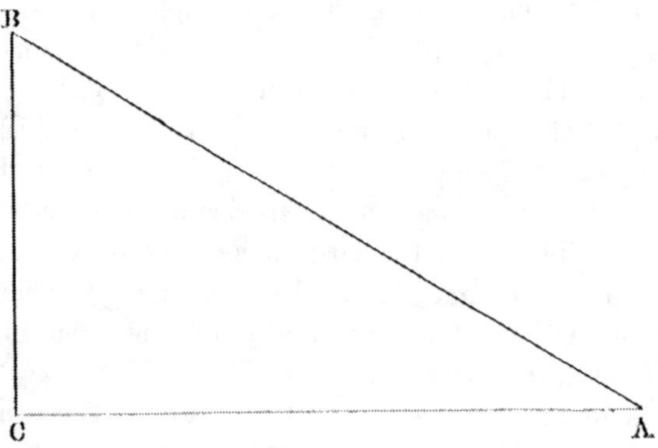

portion to the number of times C B must be taken to measure C A. In our attempt to measure the sun's distance we are limited to a base line equal in length to the *earth's* diameter, and hence it becomes necessary to employ every refinement of art to eliminate as far as possible the errors involved in the measurement of the angle C B A, or its complement, the angle C A B, on which, in the application of these principles to the problem in question, depends the measurement of the sun's distance. This quantity is the great key which unlocks all the mysteries of the entire system. Upon it depends directly the mass, volume, and density of the sun, the distances, weights, and magnitude of all the planets, and even the masses and distances of the fixed stars. It is for this reason that modern science has spared neither time nor money, neither skill nor ingenuity in the effort to reach an exact solution of this grand problem.

THE SOLAR PARALLAX.—In case an observer were located at the sun's center, and from his eye two visual rays were drawn, one to the center of the earth, the other tangent to the spherical surface of the earth, these rays would form an angle with each other at the eye of the observer, and this angle is called the *sun's horizontal parallax.*

Thus S representing the sun's center, C the center of earth, C R a radius of the earth perpendicular to the visual ray S C, and S R the visual ray drawn to the extremity R of the radius, the angle R S C is the *solar parallax*, and in case it were possible to measure that angle, as the angle S C R is a right-angle, the remaining parts of the triangle R S C become known by computation. Thus it appears that the problem of measuring the sun's distance from the earth resolves itself into obtaining the value of the *sun's horizontal parallax*, or the angle under which the earth's radius would be seen from the sun's center.

No instruments have yet been constructed sufficiently delicate to accomplish *directly* the measure of this important quantity with the requisite precision. But there is an *indirect* method, which has been employed by modern astronomers to accomplish the same object, which has been rewarded with satisfactory success. This method we shall now proceed to explain.

From the most remote antiquity it has been known that there are two planets, *Mercury* and *Venus*, which appear to revolve around the sun, never receding from that orb beyond certain narrow and well defined limits. The distances from these planets to the sun are less than the earth's distance from the same luminary, and hence they must at each of their revolutions pass between the earth and sun. Modern science has confirmed these ancient discoveries, and the telescope has even shown that on certain rare occasions each of these planets actually passes between the solar disk and the eye of an observer on the earth, and appears as a *round black spot* on the bright surface of the sun. These passages of the planets across the solar disk are called *transits*, and it happens that the *transits* of *Venus* furnish an admirable means of reducing the errors involved in the direct measurement of the solar parallax, as we shall now proceed to explain.

We will first present the principle involved, and then make the application.

Let it be required to determine the distance of the point A from any inaccessible surface, as C D, and that A A′ is the longest base line which can possibly be employed. In case the distance of the point B′ on the surface C D be required, then the angles B′ A′ A and B′ A A′ must be measured, and their sum, subtracted from 180°, gives for a remainder the angle A B′ A′, or the angle under which the line A A′ would be seen by a spectator at B′. Now this angle, because of its minute value, may be difficult to measure, and we desire to find some artifice by which this difficulty may be at least diminished, if not entirely removed. Suppose then a material point to be located at B, much nearer to A A′ than to C D, an observer at A would see the point B projected on C D at B″, while an observer at A′ would see the same point

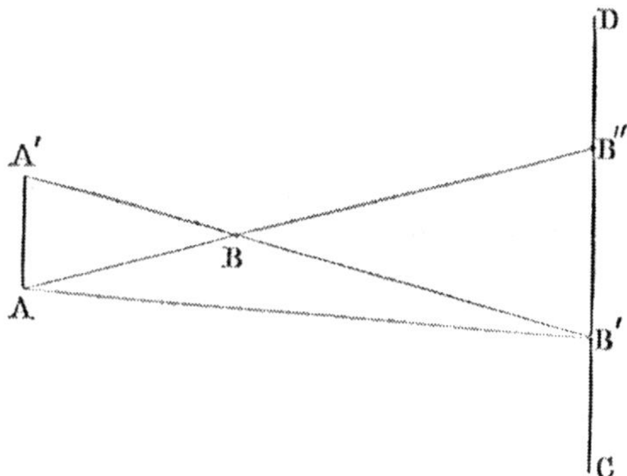

projected at B′. Now let us suppose that the points B′ and B″ can be identified and seen as round, black, permanent spots on the remote surface C D; in case B is further from C D than from A A′, it is clear that the *visual angle* subtended by B′ B″, as seen from A, will be larger than the visual angle subtended by A A′, as seen from B′, in the proportion of the distance B B′ to the distance B A′; and if B B′ should be 2½ times longer than B A′, then would B′ B″ be 2½ times longer than A A′; and the problem resolves itself into the measurement of the large angle B′ A B″ instead of the small angle A B′ A′.

Such is the principle; and we will now proceed to its application. A A′ is the diameter of the earth. B is the planet Venus, and C D is a diameter of the solar disk. To an observer at A, Venus is seen on the sun as a black spot at B″, while an observer at A′ sees the planet projected at B′. Venus is about 2½ times further from the sun than from the earth, hence B B′ is 2½ times longer than B A′, and therefore B′ B″ is 2½ times greater than A A′, or 2½ times greater than the diameter of the earth, as seen from the sun, or *five* times greater than the sun's horizontal parallax; it is therefore but one fifth part as difficult to measure the angle B′ A B″ as to measure the angle A B′ A′.

There is another important advantage gained in using the *transit of Venus* in the measurement of the solar parallax, arising from the fact that modern science has obtained a very exact knowledge of the *relative* velocity of Venus across the solar disk.

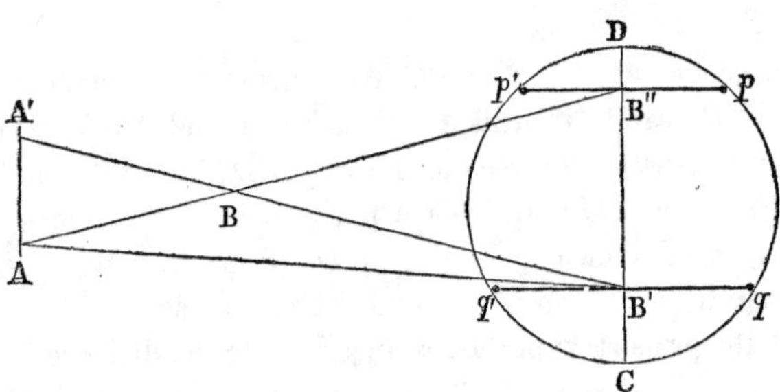

If we note, then, exactly the moment the planet is in contact with the solar disk at p, and also at p′, this interval of time will give an enlarged measure of the chord pp′, described by the planet as seen from the station A. In like manner the observer at A′ making the same observations at q and q′, we shall obtain the relative lengths of these two chords, and hence an accurate measure of the interval B′ B″, by which they are separated, or of five times the *solar parallax.*

Although this problem may appear somewhat complex at the first, careful study will render it, in these general outlines, very simple and easily intelligible. Its high value in the measurement of the very most important element in the entire system of the sun and his satellites should secure from the student all the time and attention necessary to its complete mastery.

Such is the importance attaching to this great problem, that at the last transit of Venus, governmental expeditions were fitted out at great expense, and observers were dispatched to points on the earth's surface as far asunder as possible, each observer noting, with every precaution, the exact time in hours, minutes, and seconds, from the first contact of the planet with the sun's disk, up to the moment of last contact.

It will be seen that the problem, as presented above, is freed from many complications which surround it in practice, such as those arising from the revolution of the earth in its orbit, its rotation on its axis, and the fact that the observers are not located at the extremities of the same diameter of the earth. These and other matters affecting the result being carefully taken into account, we have obtained, for the value of the sun's *horizontal* parallax, when at his mean distance from the earth, 8″.6, or eight and six-tenths seconds of arc, showing that this grand orb is removed from the earth to a distance of about ninety-five millions of miles.

We shall recur to the *transits* of Mercury and Venus when we come to treat of those bodies.

THE SUN'S LIMB

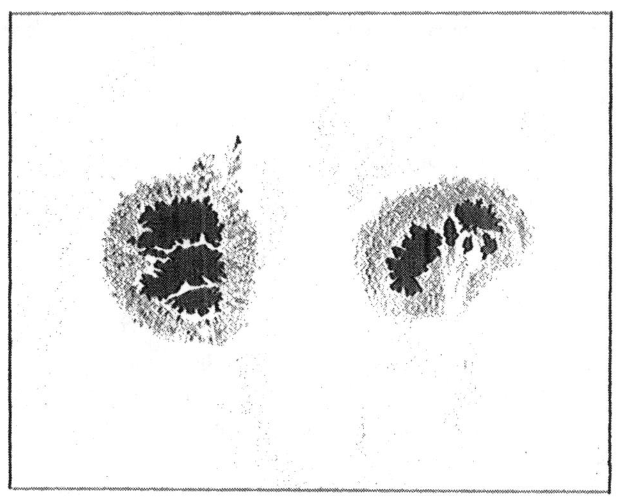

SOLAR SPOTS

THE SUN'S REAL MAGNITUDE.—Modern instruments enable us to measure with great exactitude the angle subtended by the sun's apparent diameter, an angle whose value at the sun's mean distance amounts to 32′.1″. But a globe removed to a distance of ninety-five millions of miles, and yet having an apparent diameter of 32′.1″, must have a real diameter of no less than 882,000 miles in length, or more than one hundred and eleven times longer than the diameter of our earth, as we shall hereafter see. This enables us to compare the bulk or volume of these two globes, and we find that it would require no less than one million three hundred and eighty-four thousand four hundred and seventy-two globes as large as the earth to fill the vast interior of a hollow globe as large as the sun. This is a comparison of bulk only; the relative weights of the earth and sun must be considered hereafter.

If this wonderful globe excited our admiration by the splendor of its surface, and its floods of light and heat, how must this admiration be increased when we contemplate its great distance and its gigantic proportions?

THE PHYSICAL CONSTITUTION OF THE SUN.—But for the aid derived from the telescope man could never have passed beyond mere conjecture as to what lies on the surface of the sun. The telescope, however, magnifying a thousand times, transports the observer over a vast proportion of the distance separating him from the solar orb, and plants him in space within ninety-five *thousand* miles of the sun's surface, there to examine the phenomena revealed to his sight by this magic tube. We may, therefore, regard the sun's distance as reduced to the thousandth part of its actual value, and we should not be surprised to find upon a globe of such grand proportions fluctuations and changes which, at this reduced distance, may become distinctly visible. This anticipation has not been disappointed.

THE SOLAR SPOTS.—To the naked eye the sun's surface presents a blaze of insufferable splendor, and even when this intense light is reduced by the use of any translucent medium, the entire disk appears evenly shaded, with a slight diminution of light around the circumference, but without visible spot or variation. When, however, the power of vision is increased a hundred or a thousand fold by telescopic aid, and when the intense heat of the sun and his equally intense light are reduced by the interposition of deeply colored glasses, the eye recognizes a surface of most wonderful character. Instead of finding the sun everywhere equally brilliant, the telescope shows sometimes on its surface *black spots*, of very irregular figure, jagged and broken in outline, and surrounded by a *penumbra* conforming in figure to the general outline of the central black spot (called the *nucleus*,) but of much lighter shade. Even where there are no spots, the surface of the sun is by no means uniformly brilliant. The entire surface has a *mottled* appearance. with delicate pores or points, no one of which can be readily held by the eye, but a group of them may sometimes be seized by the vision under favorable atmospheric circumstances, and can be held long enough to demonstrate that these minute pores do not change their relative position, or disappear while under the eye.

Besides the mottling of the surface, the telescope detects in the solar orb a variety of brighter streaks, called *faculæ*, whose appearance has been connected, as some believe, with the breaking out of the black spots.

Watching from day to day a single spot, or a group

of spots, on the sun's surface, they are found to advance together in the same direction, slowly to approach the edge of the sun, finally to disappear from the sight, and after a certain number of days to re-appear on the opposite side of the sun's disk, revealing the surprising fact that the sun is slowly rotating on an axis whose position seems to be invariable. In case these spots were absolutely fixed on the sun's surface, they would reveal the exact period in which his rotation is performed, but in consequence of their change of figure, and change of position as well, we can only reach an approximate value of the period of rotation. This is now fixed, by the best authorities, at *twenty-five days, eight hours* and *nine minutes.*

During the past thirty years M. Schwabe, of Dessau, has given special daily attention to counting the groups and spots on the sun, and by preserving a record it has been discovered that the amount of solar surface covered by the black spots is not only variable but that *periodicity* marks this variation. The entire change, from a *maximum* of spots counted in any year, to the *minimum*, occupies about five and a half years, and the same time elapses from a *minimum* to a *maximum*, making the period from *maximum* to *maximum eleven years.* This fact is one of the most surprising revealed in the physical constitution of any of the heavenly bodies, and thus far has baffled the power of human investigation to explain it, while its mysterious character is increased by the fact recently discovered, that this periodicity in the solar spots is identical in duration with a certain variation observed in the intensity of terrestrial magnetism. Thus, it would seem, that a new bond of union is about to be established between the earth we inhabit and that mighty orb whence we receive our supplies of light, and heat.

Some astronomers account for the solar spots by supposing the sun to be a solid, dark, opaque globe, surrounded by two atmospheres, the exterior one a highly luminous and gaseous envelope, the interior more dense, and possessing great reflecting power. The spots are supposed to result from powerful internal convulsions, upheavals from within breaking through these two envelopes, and producing a more extended chasm in the external luminous atmosphere. I have examined the surface of the sun and closely observed the large solar spots with a refractor of admirable performance, and so far from presenting an appearance such as the above hypothesis would warrant, the entire exhibition resembled the openings often found by melting through a thick stratum of solid ice from below—the spiky and jagged outline of the black nucleus being well represented by a similar form in the *opening* through the ice, while the penumbra was very faithfully represented by the *thinner portions* of ice remaining around the opening. It is not to be inferred from this comparison that the author entertains the opinion that the exterior of the sun is a solid crust, and that these solar spots are produced from the melting of this crust by the action of internal fires. The comparison is made for the purpose of illustrating, as strongly as possible, the absolute appearance of these inexplicable phenomena, and to present as strong a contrast as the facts warrant to the statement made by a distinguished astronomer, that the sun's surface, when viewed by a powerful telescope, resembles "the subsidence of some flocculent chemical precipitates in a transparent fluid." So far from this being the case, the sharp outlines of the penumbra surrounding the dark spots has often been seen to cut directly across the minute pores, dividing them sharply and sometimes equally.

Recent observations seem to demonstrate that what has generally been considered the solar surface is really the exterior of a cloudy atmosphere beneath the luminous ocean surrounding the sun. Mr. Dawes, by an eye-piece of his own construction, bearing a metallic diaphragm, in which a minute hole is pierced, coincident with the axis of the telescope, has been enabled to make a very critical examination of the solar spots. He finds in the center of the dark spot a smaller opening, which is, as now seen, *intensely black*, and this is at present regarded as the real surface of the solar orb. The same distinguished observer has announced the discovery of an actual *rotation* of the solar spots about a central axis. This important fact has given rise to speculation as to the probable cause of these wonderful fluctuations which occur in the solar atmospheres.

It is conjectured that these exhibitions may be produced by tremendous storms or whirlwinds resembling those which sometimes sweep over the surface of the earth, and whose vortices, if seen from above, would present an appearance not unlike the spots on the sun. We understand how these tornadoes are generated in the atmosphere of the earth, but it is useless to attempt to conjecture the causes which can produce such amazing effects in the solar atmosphere.

INTENSITY OF THE SOLAR HEAT.—Admitting that the heat of the sun falling on the earth is diminished in the ratio of the square of the sun's distance, it is not difficult to form some approximate idea of the intensity of the solar heat at the surface of the sun. By exposing a surface of ice to the direct action of the sun's heat, when the sun was nearly vertical, Sir John Herschel determined by experiment the thickness of the ice melted in a given time.

From this and like experiments it is determined that it would require the combustion of more than *one hundred and thirty thousand pounds* of coal per hour on each *square foot* of the sun's surface to produce a heat equal to that radiated from the solar orb.

When an image of the sun is received on any surface it is found that the central point of the image is more heated than the parts near the circumference, and that the temperature diminishes from the equator toward the poles.

THE SUN'S ATMOSPHERE.—These facts have been accounted for by supposing the sun to be surrounded by a dense atmosphere, and that the heated rays which pass through the deepest part of this atmosphere lose a portion of their heat, and hence the regions around the disk of the sun should be, to us, less heated than those near the center of the solar orb. There are some phenomena attending a total eclipse of the sun which seem to sustain this hypothesis of a solar atmosphere. At the moment the eclipse becomes total, there is seen to burst from the jet black disk of the moon a sort of *halo* or *glory*, radiating on every side, and presenting a spectacle of wonderful grandeur, so much so that on the occasion of the eclipse of July, 1842, witnessed at Pavia, the entire populace burst into a shout of wonder and admiration.

There also appeared, at the same time, *flames* of *fire* darting from behind the limb of the moon, resembling mountains of rose-colored light, rising to the height of forty or fifty thousand miles above the surface of the sun.

These *flames* are known to assume the form of cloudy exhalations which, in some instances, seem to be drifted like smoke ascending in a calm atmosphere to a certain level, where it meets a current and is borne off horizontally.

There is another phenomenon attending the rising and setting of the sun at certain seasons of the year in the shape of a vast beam of faint, gauzy light, of lenticular form, rising from the point of sunset in the evening, and stretching upward in the direction of the sun's path sometimes 70° or 80°. This is called the *Zoadical Light*, and has long been regarded as the evidence of uncondensed nebulosity, or a material atmosphere surrounding the equatorial regions of the sun. The central line, or axis, of this luminous beam does not appear to be fixed in position, and hence a difficulty arises not readily removed by the hypothesis of a material atmosphere.

Some have supposed this mysterious luminous zone to be a nebulous ring surrounding our moon, while others have regarded it as an immense ring of minute asteroids or meteors, revolving round the sun, and slowly subsiding into this grand luminary, and by the conversion of their *velocity* into *heat*, as they fall in a perpetual shower on the sun, or are burned up in the solar atmosphere, keeping up a supply equal to the vast radiation shot forth from the sun at every moment of time. While we are willing to admit that a material globe, falling into the solar atmosphere, may generate immense heat, in proportion to its magnitude and velocity, it seems quite impossible to adopt the hypothesis that the zodiacal light is either a material solar atmosphere or a ring of revolving meteors, as it extends to such a vast distance from the sun, that if revolving with the sun, as does our atmosphere with the earth, the particles would be thrown beyond the control of the sun and would be dissipated into space.

We are compelled to acknowledge that up to the present time science has rendered no satisfactory account of the origin of the solar light or heat. Whence comes the exhaustless supply, scattered so lavishly into space in every direction, we know not. Neither is it possible to give a satisfactory solution of the solar spots, or of any of the strange phenomena attending their rotation or translation on the sun's surface. The idea that tornadoes and tempests rage in the deep, luminous ocean that surrounds the sun, like those which sometimes agitate the atmosphere of the earth, has no solid foundation. We know the exciting causes of the tornadoes on earth, but why such storms should exist in the solar photosphere it is in vain to conjecture at present. Doubtless the time will come when these phenomena will be explained. Persevering and well-directed observation will, in the end, triumph; but these are matters which must be consigned to the researches of posterity.

CHAPTER II.

MERCURY, THE FIRST PLANET IN THE ORDER OF DISTANCE FROM THE SUN.

Its Early Discovery.—Difficult to be distinguished from the Stars.—Elongations.—Motion Direct and Retrograde.—Sometimes Stationary.—Nature of the Orbit.—Variation in the Elongation Explained.—The Nodes.—Transit of Mercury.—Inclination of Mercury's Orbit.—Mean Distance from the Sun.—Conjunctions.—Phases.—Diameter and Volume.

No discovery made by the ancients gives us a higher idea of the care and scrutiny with which their astronomical observations were conducted than the fact that the minute planet Mercury, so difficult to be seen, and so undistinguishable from the fixed stars, was discovered in the very earliest ages of the world. That the brighter planets, such as Venus and Jupiter, whose brilliancy exceeds that of any of the fixed stars, should have been detected to be wandering bodies, even in the remotest antiquity, is by no means surprising. For in watching the sun rising and the sun setting, so as to note, in the first instance, the stars nearest to the sun, which were the last to fade away, and in the second, those stars which were the first to become visible, the change of position of the planets Venus and Jupiter could not fail to attract the attention of the student of the heavens; but the planet Mercury is so small, and so rarely visible, even to the keenest eye, that it is said Copernicus himself, during his whole life, devoted to the study of the heavens, never once caught sight of this almost invisible world.

Mercury, in his appearance to the naked eye, is not distinguishable from the fixed stars. His close proximity to the sun, the fact that he is never visible except near the horizon, and the intense brilliancy of his disk give to him that twinkling appearance which distinguishes the fixed stars. Notwithstanding all these difficulties the oldest astronomers managed to acquire a very complete knowledge of the principal facts connected with the movement of this planet. By a careful and continuous examination it was found that Mercury never receded more than about twenty degrees from the sun's center. The amount of recess, or *elongation* as it is called, was soon discovered to be a variable quantity, a fact which demonstrated that in case the planet revolved in a circular orbit, inclosing the sun, the sun could not occupy the center of this circle. By watching the elongations from *revolution* to *revolution*, it was found that they varied from a minimum of 16° 12′, to a maximum of 20° 48′. Knowing the amount of this variation, and watching carefully the progressive change, it became possible to reach a tolerably accurate knowledge of the nature of the orbit described by the planet in its revolution around the sun. It was soon discovered that in some portions of his orbit Mercury advanced with the sun in his march among the fixed stars, while in other parts of his orbit his motion became *retrograde*, and in the change from direct to retrograde, and the reverse, the planet apparently ceased to move, and for a short time became stationary.

It will be seen that all these changes are readily accounted for by supposing the planet to revolve about the sun in a circular orbit, the sun being eccentrically placed. If we conceive two visual rays, to be drawn from the eye of the observer, and tangent to the orbit of Mercury on the right and on the left, the planet, while traversing that arc of its orbit intercepted between the points of contact and nearest to the eye, will move *direct;* in passing through the point of contact after direct motion ceases, it will move off in the direction of the visual ray, and hence will appear *stationary* for a short time. In the larger portion of its orbit (that remote from the eye) its motion must be opposite to that of the sun, and hence *retrograde.* In coming up to the second point of contact the planet will move along the visual ray toward the eye of the observer, and hence for a short time will appear stationary.

To account for the variation in the elongations of Mercury, we must either suppose the point of nearest approach of the planet to the sun, called its *perihelion,* to be in motion, or else we must suppose the spectator to be himself moving, and thus to behold the planet, its perihelion point, and the sun, under varying relations to each other. As the early astronomers assumed the immobility of the earth, they explained the variations in the elongations of Mercury by giving to its perihelion point a motion of revolution about the sun.

It is impossible to follow the planet with the naked eye in its close approach to the solar orb, as its feeble reflected light is necessarily overpowered by the brilliancy of the sun, but by close observation, and by marking the positions of the planet at its disappearance and reappearance, the old astronomers are said to have reached to a knowledge of the fact that this planet sometimes crosses the sun's disk, producing what is called a *transit of Mercury,* identical in its phenomena with the transit of Venus, already spoken of in connection with the determination of the solar parallax. In case the plane of the orbit of Mercury were exactly coincident with the plane of the sun's apparent orbit, it is manifest that every revolution of the planet would produce a transit. As this, however, is not the case, and as no central transit can occur, except when the planet crosses the visual ray drawn from the eye of the observer to the sun's center, it is manifest that the planet Mercury, during a central transit, must actually pass *through* the *ecliptic* from one side of this plane to the other. This point of passage through the plane of the sun's apparent orbit is called the *node* of the planet's orbit. There are, of course, two such points. The planet passes its *descending* node in moving from the north to the south side of the ecliptic, and its *ascending* node on its return from the south to the north side.

It is thus seen that in order to produce a transit of Mercury there must be *a conjunction* of the planet, its node, and the sun. Whenever this conjunction is absolute, Mercury will pass across the sun's center. When it is only approximate, the planet will transit a small portion of the sun's disk, or possibly pass without contact at all.

An attentive examination of the places of the planet, before and after a transit, led to a pretty accurate determination of the angle under which the plane of the planet's orbit is inclined to the plane of the ecliptic. This angle was *approximately* determined by the ancients, while *modern* science fixed it at the commencement of the present century at 7°.00′.10″.

The motion of Mercury in its orbit is more rapid than that of any of the planets thus far discovered, traveling, as it does, more than one hundred thousand miles an hour, and performing its entire revolution about the sun in about eighty-eight of our days. In case this world has the same variety of seasons which mark the surface of our own earth, these will follow each other in such rapid succession that the longest of them will consist of only about three of our weeks. It is not difficult to compute the intensity of solar light and heat which falls upon the surface of the planet Mercury, in case these be subjected to the same modifying influences which exist upon the earth. But as we remain in ignorance of the circumstances which surround this distant planet, it is vain to speculate upon the physical constitution of a world whose close proximity to the sun has thus far shut it out from the reach of telescopic examination.

The distance of the planet Mercury from the sun may be readily determined, in certain portions of its orbit, in case we know first the earth's distance from the same orb. For example, conceive a visual ray to be drawn from the earth, tangent to the orbit of Mercury (supposed, for the present, to be circular); place the planet at the point of contact, and join the center of the planet with the center of the sun; also join the centers of the earth and sun—the triangle thus formed, having the earth, Mercury, and the sun as the vertices of its three angles is right-angled at Mercury, while the angle at the earth is readily measured, and is nothing more, indeed, than the elongation, for the time being, of that planet. Hence, in the right angled-triangle, we know the angles and the longest side, extending from the earth to the sun, and by the simplest principles of trigonometry, we can compute the remaining parts—namely, the distance of Mercury from the sun and from the earth. By this, and by other methods more accurate, it is found

3

that Mercury revolves in an orbit around the sun, and at a mean distance of about *thirty-six millions of miles.*

As the entire orbit of this planet lies within the limits already assigned, it follows that the planet can never be seen in a quarter of the heavens opposite to the sun, or can never be in *opposition.* When nearest the earth, and on the right line joining the sun and earth, Mercury is said to be in *inferior* conjunction. When 180° distant from this place it is on the other side of the sun, with respect to the earth, and is then in its *superior* conjunction.

The telescope has demonstrated that this planet passes through changes like those presented by the moon. When in superior conjunction the planet will be seen nearly round, as in that position nearly the whole of the illuminated surface is turned toward the eye of the observer on the earth. As the planet comes round to its inferior conjunction the light gradually wanes, until at inferior conjunction a slender crescent of great delicacy and beauty is revealed to the eye, provided the planet does not lose its light entirely in the passage across the sun's disk. These *phases* of Mercury prove, beyond question, the fact that the planet does not shine by its own light, but that its brilliancy is derived from reflecting the light of the solar orb.

The degree of precision reached in predicting the transits of Mercury indicates, with wonderful force, the progress of modern astronomy. The first predicted transit which was actually observed occurred in 1631, when the limits of possible error were fixed by the computer at *four days;* and hence the watch commenced two entire days before the predicted time.

If the transit had taken place in the night time, the

opportunity for verification would have been lost. Fortunately this was not the case, and the toil and zeal of Gassendi were rewarded with the first view of Mercury projected on the solar disk ever witnessed by mortal man. Nearly two hundred years later, at the beginning of the nineteenth century, the French astronomers ventured to assert that their predictions could not be in error more than *forty minutes.* The transit which occurred on the 8th Nov., 1802, verified this assertion very nearly. By a more careful study of the causes affecting the place of the planet, forty-three years later, the discrepancy between computation and observation was reduced to only *sixteen seconds* of time, a quantity very minute, when we take into account the variety of causes affecting the resolution of the problem. The transits of Mercury recur at certain regular intervals, repeating themselves after a cycle of 217 years, falling for the present in the months of May and November.

Having learned the distance of Mercury from the earth, and having measured the angle subtended by its diameter, we find its actual magnitude to be much smaller than that of the earth. Its diameter is but 3,140 miles, and its volume is but 0.063, the earth's volume being counted as unity.

In comparison with the vast proportions of the sun, this little planet sinks into absolute insignificance, for if the sun be divided into a million equal parts Mercury would not weigh as much as the half of one of these parts.

CHAPTER III.

VENUS, THE SECOND PLANET IN THE ORDER OF DISTANCE FROM THE SUN.

THE FIRST PLANET DISCOVERED.—MODE OF ITS DISCOVERY.—HER ELONGATIONS. —MORNING AND EVENING STAR.—A SATELLITE OF THE SUN.—HER SUPERIOR AND INFERIOR CONJUNCTIONS.—HER STATIONS.—DIRECT AND RETROGRADE MOTIONS.—THESE PHENOMENA INDICATE A MOTION OF THE EARTH.—TRANSITS OF VENUS.—INCLINATION OF THE ORBIT OF VENUS TO THE ECLIPTIC.—HER NODES.—INTERVALS OF HER TRANSITS.—KNOWLEDGE OF THE ANCIENTS.—PHASES OF VENUS.—HER ELONGATIONS UNEQUAL.—NO SATELLITE YET DISCOVERED.—SUN'S LIGHT AND HEAT AT VENUS.—HER ATMOSPHERE.

THIS planet is the second in order of distance from the sun, and as it is the most brilliant of all the orbs, with the exception of the sun and moon, it was undoubtedly the first discovered of all the planets. The movements of the sun and moon among the fixed stars must have claimed the attention of the observers of celestial phenomena in the earliest ages of the world. In marking the rising and setting sun, and in noting the stars which were the last to fade out in the morning twilight and the first to appear in the evening after the setting of the sun, the brilliancy of Venus could not fail to have attracted the attention of the very first observer of celestial phenomena. A star of unusual brightness was noticed in comparative proximity to the sun in the early evening. The sun's place, with reference to this object, having been carefully marked, for a few consecutive nights, it was found that the distance between them was rapidly diminishing. It was readily seen that this diminution of distance was due to the fact that the bright star was approaching the sun, for by comparing its place among the fixed stars with what it was a few nights previous, this star was found to have changed its position among the group in which it happened to be located, and was evidently advancing rapidly toward the sun.

We are thus presented with the exact facts which must have marked the discovery of the *first planet* or wandering star ever revealed to the eye of man. We know not the name of the discoverer, nor the age or nation to which he belonged, but we are satisfied that the facts as above stated did undoubtedly occur; and we find not only profane authors but one of the Hebrew prophets referring to this planet more than two thousand five hundred years ago. The student who desires may easily re-discover the planet Venus. She will be readily recognized as the largest and brightest of all the stars, and will be found never to recede from the sun more than about 47°. From this distance, which she reaches at her greatest elongation, the planet will be found, at first slowly, but afterward more rapidly, to approach the solar orb. She will finally be lost in the superior effulgence of the sun; and when the unaided eye ceases to follow her in her approach to the sun, telescopic power will enable the observer to continue his observations until, finally, the sun's direct beams, mingling with those of the planet, she ceases to be visible, and is now lost for a greater or less period, until she emerges from the solar rays, appearing just before the sun in the gray morning twilight. She now recedes from her central orb, finally reaches her greatest elongation upon the opposite side, stops in her career, returns again, and thus oscillates backward and forward, never passing certain prescribed limits.

As already stated, the fact that Venus was a planet or wandering star must have become known among the very first of astronomical discoveries; but it required, doubtless, a long series of observations to determine the truth that the bright star, which for some months had accompanied the setting sun, and which was at length lost in the solar beams, was the same object which, at a later period, became visible in the morning dawn, having passed by or across the solar disk. This discovery, however, is said to have been made by the Egyptian priests, and was by them communicated to the Greek astronomer, Pythagoras, who taught this truth to his countrymen.

It is obvious, from the above facts, that the planet Venus, like Mercury, is beyond doubt a true satellite of the sun, even to the inhabitants of the earth, and it is equally manifest that, whatever be the true relations between the earth and the sun, and whichever one of these two bodies may be at rest, one thing is certain, the planets Mercury and Venus cannot by any possibility have the earth for their center of motion. No matter in what region of the heavens the sun may be found at any season of the year, these two interior planets ever accompany him. As Venus recedes to a greater distance from the sun than Mercury, it follows that her orbit of revolution around the sun must be the larger of the two. We are thus enabled, by the simplest train of reason and observation, to fix the following facts:—The sun is a central orb, about which revolve, in regular order, two planets, the nearest of which is Mercury, and next to Mercury, Venus, with periods of revolution, readily determined by the spectator on the earth's surface. These facts are exceedingly important as the primary ones which lead to the discovery of the true system of the universe.

When Venus passes between the eye of the observer and the sun, she is said to be in her *inferior* conjunction; when she is directly beyond the sun, with reference to the spectator, she is in her *superior* conjunction. From her inferior to her superior conjunction she occupies a position west of the sun, rises in the early morning, before the sun, and is known as Phosphorous, or Lucifer, or the morning star. From her superior to her inferior conjunction she follows the setting sun; she becomes our evening star, under the name of Hesperus.

In examining the phenomena involved in the motions of Venus, and watching her carefully in her approach to and in her recess from the sun, it is found that her movements are almost identical with those of Mercury—her motions for a certain portion of her revolution being *direct*, or like those of the sun; she then becomes stationary, then moves backward or *retrograde* among the fixed stars, becomes stationary again,, and then commences her direct movement. All these facts are readily accounted for by admitting that Venus revolves about the sun in an orbit nearly circular, and that she is viewed by a spectator situated exterior to her orbit, and moving around the sun and Venus in a circle, whose plane makes a small angle with the plane on which the orbit of Venus lies. If a visual ray be drawn from the eye of the observer, tangent to the orbit of Venus, should the planet happen to fill the point of contact, she will appear to move in the direction of this ray, and, for the time being, will be directly advancing to, or receding from, the eye of the observer, and thus will appear stationary. That the observer is in motion, is manifest from the fact that the direct movement of Venus does not bear that relation to the retrograde movement which is required by such an hypothesis. Indeed, if two visual rays were drawn from the eye of a stationary observer, tangent to the orbit of Venus, she would appear to move from one point of contact to the other, on the hither side of her orbit, with a direct motion, while on the further side of her orbit, between the points of contact, her motion would appear retrograde. These facts, however, are not presented in nature, and would be subverted, of course, by supposing the spectator to be in motion. In case the spectator were to occupy the line passing from the sun's centre through Venus, and to revolve about the sun in the same period occupied by the planet, then would the planet always be seen in inferior conjunction with the sun. As this is not the fact in observation, it is manifest that the angular velocity of the spectator is not so great as that of the planet Venus, as she finally emerges from the sun's rays, after her inferior conjunction, beyond the line, joining the sun's center and the eye of the beholder. Here, then, is another important fact, which must be taken into account when we shall inquire into the true system of nature, as presented in the organization of the planetary worlds.

In case the eye of the observer were located in the same plane in which the orbit of Venus lies, this plane, passing, as it does, through the sun's center, it is clear that at every inferior conjunction of the planet there might be seen *a transit of Venus*, while at every superior conjunction, the planet would be *occulted*, or hidden, by passing actually behind the disk of the sun. It happens, however, that the plane of the orbit of Venus does not coincide with the plane of the ecliptic, or earth's orbit. These planes are inclined to each other, under an angle of 3° 23′ 28″.5, one half of the orbit of Venus lying above, or north of the ecliptic, the other half lying below, or south of the ecliptic. The point in which Venus passes from the north to the south side of the ecliptic is called the *descending node.* She returns from the south to the north of this plane through the *ascending node*, and the line joining these two points is called the *line* of *nodes.*

The transits of Venus, unfortunately for astronomical science, are of very rare occurrence, and are separated by intervals of time which are very unequal. The periods from transit to transit are 8,122, 8,105, 8,122, &c., years, for a long period falling in the months of June and December. As already stated, no transit can occur except when the planet is in the act of passing her node at her inferior conjunction, while, at the same time, the earth is crossing the line of nodes of the planet prolonged. This line of nodes, though not fixed, moves very slowly, and at this time crosses the earth's orbit in those regions passed over by the earth in the months of June and December. After a transit the relative motion of Venus, the earth, and the node of the orbit of Venus, is such as to render it certain that within eight years another transit will occur, as within this period Venus does not, at her inferior conjunction, recede too far from the plane of the ecliptic to render her transit impossible.

In our account of the determination of the solar parallax (Chap. I) we have stated that the distance of Venus is readily determined by the measure of her horizontal parallax. Her distance may also be determined, after we have learned the distance of the sun, by the same method used in measuring the distance of Mercury (Chap. II).

By these and other methods the mean distance of this planet from the sun is found to be about sixty-eight millions of miles, and from the measure of her apparent diameter we conclude her actual diameter to be 7,700 miles, or a little less than the diameter of the earth, as we shall see hereafter.

The period of rotation of Venus has not been well determined, but from an examination of indistinct spots, sometimes visible on her face, it is conjectured that she rotates on her axis in about twenty-four hours, or in the same period occupied by the earth.

The changes in the brilliancy of the planet Venus are accounted for in a two-fold way. In case the observer is really exterior to her orbit, as the planet's distance from the sun is on the average 68,000,000 of miles, then when the planet occupies that point in her orbit nearest the observer she will be closer to the eye than when in the opposite point of her orbit by an amount equal to no less than double her mean distance from the sun, or 136,000,000 of miles. We readily perceive that this vast increase of distance must diminish in direct proportion the apparent diameter of the planet, and thus her brightness must decline, as she recedes from her nearest to her greatest distance from the observer. To this cause, however, of a change of brilliancy, is to be added another of still greater importance. We have already stated that the planet Venus, when seen projected upon the sun's disk during her transit, appears as a round, black spot on the brilliant surface of the sun. This fact demonstrates, beyond a doubt, that the planet Venus is a dark, opaque globe, destitute of light, and only visible by reflecting the light which it receives from the sun. If further evidence of this statement were wanting, it is found in the fact that after the planet passes her inferior conjunction and becomes visi-

PHASES OF VENUS

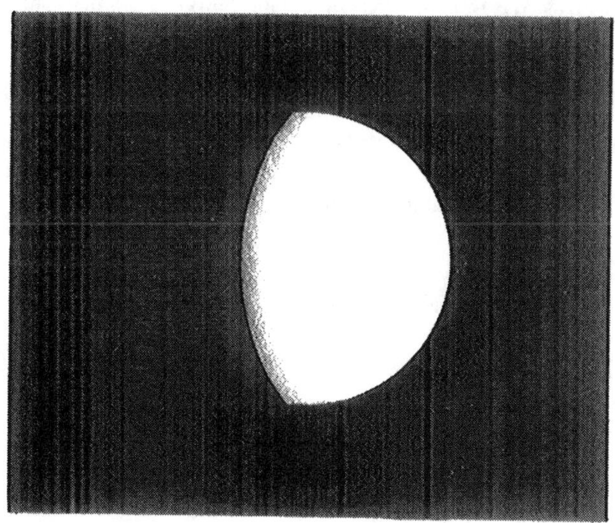

ble in emerging from the sun's beams, she is first seen by the telescope as a slender and delicate crescent of silver light. As she recedes from the sun this phase gradually changes; more and more of her illuminated hemisphere becomes visible, until, finally, at her superior conjunction, her disk becomes round and well-defined. The same facts are true of the planet Mercury, and thus is added another powerful evidence that these two planets are satellites of the sun, revolving about this luminary in orbits nearly circular, and deriving their light from this great central body.

When we come to measure accurately the greatest elongations of Venus we find them *unequal.* In case the spectator were stationary, and admitting the circular form of the orbit of Venus, these inequalities could not occur. We thus are led to believe, either that the orbit of the planet is not circular, or, if it be circular, that the sun is eccentrically situated, or that the observer himself is in motion.

It is possible that any two, or even all of these causes, may combine to produce the phenomena presented in the movements of Venus. We shall recur again to these matters when we come to consider the great problem of the true system of the universe.

The extreme brightness of this planet makes it a very beautiful but difficult object for telescopic observation. Although spots have been seen upon the surface of Venus, and by their close examination her period of rotation upon her axis has been approximately determined, I have never been able, at any time, with the powerful refractor of the Cincinnati Observatory, to mark any well-defined differences in the illumination of her surface. If we are to trust to the observations of others, the inequalities which diversify the planet Venus far exceed in grandeur those found upon our earth. It is stated by Mr. Schroter that, from his own observations, the mountains of Venus reach an altitude five or six times greater than the loftiest mountains of our own globe.

It has been affirmed by several distinguished astronomers that this planet is accompanied by a minute satellite, but by the application of the most powerful telescopes, during the present century, and after the most rigid examination, this statement has not been confirmed. It was supposed that during the transit which occurred in 1769 the disputed question as to the existence of a moon of Venus would be positively settled. While the planet was distinctly seen as a dark spot upon the surface of the sun, no telescopic power could detect any dark object which might be a satellite. Although we cannot absolutely affirm that Venus has no satellite, we may safely say, that if there be one it yet remains to be discovered.

The amount of light and heat which the earth would receive from the sun, if revolving in the orbit of Venus, would be nearly twice as great as that now received; but this does not justify us in concluding that the planet Venus has a mean temperature nearly double that of the earth. We know that a powerful influence is exerted by the earth's atmosphere to modify the solar heat. There may exist an atmosphere surrounding Venus such that the temperature at her surface may be no greater than our own. It is useless, however, as we have already remarked, for us to speculate about matters concerning which we positively know nothing. There are some indications in the telescopic appearance of Venus that she is surrounded by an extended atmosphere. When presenting the form of a crescent of light, the slender horns are found sometimes to extend beyond the limits of a semi-circumference—a fact only to be accounted for, so far as we know, by admitting atmospheric refraction.

CHAPTER IV.

THE EARTH AND ITS SATELLITE: THE THIRD PLANET IN THE ORDER OF DISTANCE FROM THE SUN.

The Earth the apparent Center of Motion.—To all the Senses it is at Rest.—The Center of the Motions of the Sun and Moon.—Explanation of the Acceleration of the Orbitual Motion of the Sun and Moon.—Ptolemy's Epicycles.—The Explanation of Copernicus.—The Sun the Center of Planetary Motion.—The Earth One of the Planets.—Objections to this Hypothesis.—The Answer.—System of Jupiter discovered by the Telescope.—The Old System superseded by the New.—The Figure and Magnitude of the Earth.—How determined.—The Earth's Motions.—Rotation and Revolution.—A Unit of Time furnished by the Earth's Period of Rotation.—Earth's Orbitual Motion.—Vernal Equinox.—Perihelion of Earth's Orbit.—Its Period of Revolution.—Solar and Sidereal Time.

THE MOON.—Revolution in her Orbit.—Her Phases.—Earth's Line.—Eccentricity of her Orbit.—Revolution of her Apogee.—Inclination of her Orbit.—Moon's Parallax and Distance.—Her Physical Constitution.—Center of Gravity and Center of Figure.

The ancients did not reckon the earth as one of the planetary orbs. There seemed to be no analogy between the world which we inhabit, with its dark, opaque, and diversified surface, and those brilliant planets which pursued their mysterious journey among the stars. Sunk as they were, so deep in space, it was very difficult to reach any correct knowledge of their absolute magnitude. The earth seemed, to the senses of man, vastly larger than any or all of these revolving worlds. About the earth, as a fixed center, the whole concave of the heavens, with all its starry constellations, appeared to revolve, producing the alternations of day and night. It was not unnatural, therefore, knowing the central position of the earth with reference to the fixed stars, to assume its central position with reference to the sun, and moon, and planetary worlds.

There is no problem perhaps so difficult as that presented in the attempt to discriminate between *real* and *apparent* motion. To all the senses the earth appeared to be absolutely at rest. It could not be affirmed that any one had ever seen it move, or felt it move, or heard it move, while the sense of sight bore the most positive testimony to the motion of the surrounding orbs. It must be remembered that, in the primitive ages, the great objects of observation and study were the sun and moon. Five planets were indeed discovered, at a period so remote that no historic record of the facts of their discovery now exists. They seem to have been known to all the nations of antiquity, and a knowledge of their existence appears to have been derived from a common origin, as we shall have occasion to notice more particularly hereafter. A few of the more obvious phenomena presented in the planetary movements were known and studied by the old astronomers, but when these motions became to them inexplicable, they frankly confessed that these matters must be left for the study and development of posterity.

If, then, we confine our attention principally to an examination of the solar and lunar motions, and to the general revolution of the sphere of the fixed stars, in our efforts to determine the true position and condition of the earth, we shall find ourselves compelled, as were the celebrated Greek astronomers, Hipparchus and Ptolemy, to admit not only the earth's central position, but also its absolute immobility. It is, undoubtedly, central to the moon's motions, and it is equally central to the sun's movement; that is to say, all the phenomena of the solar motions are as well accounted for by supposing the earth to be the center about which the sun revolves, as by supposing the converse hypothesis, that the sun is the center about which the earth revolves.

So far, then, as these two great luminaries are concerned, the hypothesis of the earth's central position is well sustained, and almost indisputable. It is only when we extend our investigations to the inferior and superior planets, and gather together a multitude of facts and phenomena demanding explanation, that we find ourselves necessarily driven into so great complexity by retaining the central position of the earth, that at last we begin to doubt. We have already noticed the remarkable movements of the two planets Venus and Mercury. We shall find hereafter that phenomena of a like character were presented in the movements of Mars, Jupiter, and Saturn, each of which planets was distinguished by its *stations*, *retrogradations*, and *advances* among the fixed stars. The ancients not only adopted the hypothesis of the earth's central position and immobility, but, for evident reasons, likewise adopted the hypothesis that all motion was performed in circular orbits, and with uniform velocity. We have already seen, in our examination of the solar motions, that this orb did not move to the eye with uniform velocity, but this apparent deviation from uniformity was readily accounted for by supposing the earth to be placed a little eccentric with reference to the sun's circular orbit. The same facts becoming known with reference to the moon's motion, a like hypothesis was adopted, and the earth was placed eccentrically within the lunar orbit. In marking the planetary movements, they were found, however, to differ radically in some particulars from the movements of the sun and moon. While these great luminaries always advanced in their revolution among the fixed stars, the planets were found, in making their revolution, not only to stop, but for a time actually to turn back, then stop again, and finally to resume their onward movement. No eccentric position of the earth could account for these stations and retrogradations; but a very simple expedient was devised, which rendered a satisfactory account, in the primitive astronomical ages, of these curious phenomena. Retaining the central position of the earth and the circular figure of the planetary orbits, each planet was supposed to revolve on the circumference of a *small circle*, whose center was carried uniformly around on the circumference of the great circle constituting the orbit of the planet. By such machinery it will be seen that it became possible to render a satisfactory account of the stations and retrogradations of the planets, for while the planet was describing that portion of the small circle in which it revolved, nearest to the eye of the spectator, it would seem to move backward in the order of the fixed stars. Again, in coming directly toward the eye of the spectator, or in moving in the opposite direction along two visual rays, drawn tangent to its small circle, the planet would appear stationary. Such was the general exposition of the Greek astronomer Hipparchus, whose theory was enlarged and extended by his successor Ptolemy, whose theory of astronomy, based upon the central position of the earth, known as the Ptolemaic System, endured for more than fifteen hundred years. It was only after a long lapse of time, and by the discovery of a large number of irregularities in the solar, lunar, and planetary motions, making it necessary (to render a just account of them) to increase the number of these small circles, which were called *epicycles*, that the whole scheme finally became so cumbrous and complicated that, after long and laborious study, extending through more than thirty years of diligent observation, the great Polish astronomer, Copernicus, found himself compelled to abandon the old hypothesis of the central position of the earth, and to attempt a new solution of the great problem of the universe.

In giving up the earth as the centre about which the worlds were revolving, there was little difficulty in selecting the object which, in greatest probability, occupied the true center. All the movements of the sun could, without the slightest difficulty, be transferred to the earth, and thus the sun could become central to the earth, revolving as one among the planets. This hypothesis did not require any change whatever in the computation of those tables which gave from day to day the sun's apparent place among the fixed stars. Again, as we have already seen, the planets Mercury and Venus were undoubtedly satellites of the sun, whether the sun be at rest or in motion; and with these suggestions, the vigorous mind of Copernicus, transferring himself, in imagination, to the sun, and thence looking out upon the planetary revolutions, found that a large number of those complexities and irregularities which had so confounded him when viewed from the earth's surface were swept away for ever. When seen from the sun, as the center of motion, all the stations and retrogradations in the planetary revolutions disappeared. The complications in the movements of Mercury and Venus were reduced to perfect order and simplicity when seen from the sun. The earth itself assumed its proper rank among the planetary worlds, dignified by the attendance of its satellite the moon, and beyond the earth, the planets Mars, Jupiter, and Saturn, performed their orderly revolution in orbits nearly circular. Such is the true scheme of nature in its grand outlines, as given to the world by Copernicus. It will be seen that one of the remarkable features of the old system, namely, the uniform circular movement of the planets, was retained by the Polish astronomer. By the use of eccentrics and epicycles, Copernicus found it possible to render a satisfactory account of all the phenomena of the solar system known during his age. We can readily comprehend that a system involving the startling doctrine of the swift rotation of the earth upon its axis, and the rapid flight of its entire mass, with all its continents, and oceans, and mountains, through space, must have been received by the human mind with the greatest distrust. Indeed, there seemed to be to the eye positive proof that this bold theory was absolutely false. It was urged by the anti-Copernicans, that in case the earth did revolve about the sun in an orbit of nearly two hundred millions of miles in diameter, that the point where the axis of rotation, prolonged to the sphere of the fixed stars, pierced the heavens, must by necessity travel around and describe a curve among the stars identical with that described by the earth in revolving about the sun. Now, as no such motion of the north polar point was visible to the eye, but as the axis of the heavens remained for ever fixed among the stars, it proved beyond dispute the absolute impossibility of the earth's revolution about the sun. This train of reasoning was undeniably true, and the only response which the Copernicans could make was this: "The earth does revolve about the sun; the earth's axis prolonged does pierce the celestial concave in successive points, describing a curve precisely like the earth's orbit, and whose diameter is indeed nearly 200,000,000 of miles; but that the distance of the fixed stars is so great, that an object having this immense diameter actually shrinks into an invisible point, on account of the almost infinite distance to which it is removed from the eye of the beholder;" and with this answer the world was compelled to rest satisfied for more than two hundred years.

The doctrines of Copernicus gained a great accession of strength by the invention of the telescope. By the use of this extraordinary instrument not only were the phases of Mercury and Venus detected, but also the greater discovery of the satellites of Jupiter, presenting, in this central orb, with his four revolving moons, a sort of miniature likeness of the grander system, having the sun for its center. The simplicity of the hypothesis presented in the Copernican system, the numerous complications which it removed from the heavens, and the satisfactory account which it yielded of the discoveries made by the telescope, caused it to be adopted and defended by some of the best minds of the age immediately following that of Copernicus, among whom none is more distinguished than the great Florentine astronomer and philosopher Galileo Galillei. It is hardly necessary to mention the historical fact, that the old system of astronomy, which had held its sway over the human mind for more than 2,000 years, did not fall without a severe struggle. The astronomy of Ptolemy, and the philosophy of Aristotle, had taken so deep a hold of mankind, and were so firmly interwoven with all the systems of education and of science, that we must behold with astonishment the downfall of systems venerable from their antiquity, and whose ruin could only be accomplished by the desertion of their adherents.

THE FIGURE AND MAGNITUDE OF THE EARTH.—A knowledge of the globular figure of the earth seems to have been reached at an early period in the history of astronomy. Indeed, the concave heavens, presenting to the eye a hemisphere above the horizon, and, undoubtedly, extending beneath the earth, so as to complete the grand hollow sphere, suggested at once that the inclosed earth, minute in its dimensions when compared with the celestial globe by which it was encompassed, might also have the globular form. The curvature of the earth's surface becomes at once visible to the eye in marking the gradual approach of a ship at sea. At first only the top of the mast can be discovered, even with a glass, all the remaining parts of the vessel being hidden by the outline of the the interposed water. As the distance diminishes, more and more of the ship lifts itself above the horizon, until, finally, the water-line comes into sight. The same evidence of the rotundity of the earth is furnished by the *circular form* of the horizon which always sweeps round a beholder who ascends to the summit of a lofty mountain. Thus, we are disposed to adopt the spherical form of the earth in consequence of its simplicity, even before we have any conclusive demonstration as to its real form.

The Greek astronomers comprehended the simple process, whereby not only the true figure of the earth might be obtained, but in case it were spherical, whereby its real diameter and absolute magnitude might be determined.

This process is remarkably simple. Suppose an observer, provided with the means of directing a telescope precisely to the zenith of any given station, and in the zenith point he marks a star, which from its magnitude and position he can readily find again. Now, leaving this first station, and moving due north, measuring the distance over which he passes, he will find that, as he progresses toward the north, the star under examination will leave the zenith and slowly decline toward the south. Suppose the observer to halt, set up his instrument, and find that his star has declined one degree from the zenith toward the south.

This demonstrates that he has traveled from the first station to the second, over one degree of a great circle of the earth, or one part in 360 of the entire circumference of the earth. It follows that, in case the earth is really globular in form, the distance between the stations, multiplied by 360, will give the length of the entire circumference, and this quantity, divided by 3.14159 (the ratio between the circumference of a circle and its diameter), will give the value of the earth's diameter.

It is by methods analogous to the above that the true figure and actual magnitude of the earth have been determined. Very numerous and delicate measures, performed in many parts of the earth's surface, have revealed the surprising fact that the true figure of the earth is not that of a sphere, but of a *spheroid*, being more flattened at the poles and more protuberant at the equator than a true sphere. We shall hereafter exhibit the cause of this remarkable fact, and present some very curious and surprising results and phenomena which flow from it. By the most reliable measure we find the polar diameter of the earth to be 7,898 miles, while the diameter of the equator reaches to 7,924, being an excess of no less than twenty-six miles, which excess would have to be trimmed off to reduce the earth to a globular form.

THE EARTH'S MOTION.—We have already noticed the fact that the sun, as well as the planets thus far described, have a motion of rotation about a fixed axis, while the planets have also a motion of revolution in their orbits. Since we are compelled to recognize the earth as one of the planets, we naturally conclude that it will be distinguished by the same motions which mark the ongoings of the other planets. We shall find, indeed, that the earth has *three* motions: a motion of *rotation* about an axis, accomplished in a period of twenty-four hours, and producing an apparent revolution of the sphere of the fixed stars in the same period. A motion of *revolution* in an orbit whereby the earth is carried entirely around the sun, effecting all those changes which mark upon the earth's surface the seasons of the year, and producing at the same time an apparent revolution of the sun in a circular orbit among the fixed stars. The earth has a *third* motion, (which we will examine more fully hereafter,) occasioned by the fact that its axis of rotation does not remain constantly parallel to itself.

THE EARTH'S ROTATION.—Let us return to the consideration of the *diurnal revolution*, to the inhabitants of the earth, as well as to the student of astronomy, by far the most important motion which has been revealed by human investigation. It is, perhaps, impossible for the mind of man to form any just notion of what we call *time*, except as its flow is measured by some absolutely uniform succession of events. This perfect measure of time is found in the uniform rotation of the earth upon its axis, whereby all the fixed stars appear to the eye to perform revolutions in circles of greater or less diameter, all in the same identical period, and with a motion which, so far as we know, is absolutely uniform. Thus the duration of one rotation of the earth upon its axis, whereby any given fixed star revolves from the meridian of any place entirely round to the same meridian again, furnishes to man *a unit of time*, which, by its sub-divisions and multiplications, renders it possible to take account of historic and other events, and to mark their relations to each other, not only in the order of time, but also in the interval of time. Thus, a day is sub-divided into hours, minutes and seconds, and the fraction of a second, and by successive additions gives us larger portions of time, as weeks, months, years, and centuries. To serve this very important purpose, and to become a true unit of measure of time, it is absolutely indispensable that the motion of rotation of the earth upon its axis shall be rigorously uniform and invariable.

We have, at present, in all the active observatories in the world, a constantly accumulating power of evidence that the earth now revolves with uniform velocity. Not a star passes the meridian wire of a fixed telescope, true to the predicted moment of transit, without testifying to the absolute uniformity of the earth's rotation. So far, then, as it be possible, by human observation and human means, to determine any truth whatever, we are able to affirm the absolute uniformity of the rotation of the earth upon its axis. This truth is affirmed as of to-day; and so far as we can go back in the history of accurate astronomical observation, the same truth is affirmed of the past; and La Place informs us that, from a rigorous investigation of the whole subject, he discovers that the period of rotation of the earth upon its axis has not changed by the *hundredth part of one second of time* in a period of more than two thousand years. We will explain hereafter the train of reasoning by which this conclusion has been reached. We shall, for the present, accept the statement as a fact.

THE REVOLUTION OF THE EARTH IN ITS ORBIT.—In the examination already made of the sun's apparent revolution among the fixed stars, we have found that the revolution was performed in the same plane, cutting out of the sphere of the fixed stars an exact great circle. All that was then affirmed, with reference to the sun's *apparent* motion, must now be affirmed as belonging to the earth's *real* motion.

The earth, then, revolves around the sun in the plane of the ecliptic, at a mean *distance* of about *ninety-five millions of miles*, and in a *period* of about *three hundred and sixty-five days and a quarter.* It, of course, always occupies a position distant from the sun's place one half a circumference, or one hundred and eighty degrees. The changes of the sun's position at noon in the course of the year, which we have already examined, are now readily accounted for by the fact that the earth's axis of rotation neither coincides with the plane of the ecliptic nor is perpendicular to it, but is inclined under an angle, which is readily measured, and which is found to undergo a very slow change from century to century. In case the earth's axis were perpendicular to the plane of the ecliptic, then the illuminated hemisphere of the earth would always be bounded by a meridian circle, and every inhabitant of the earth would find his days and nights precisely equal, no matter what his location upon the earth's surface. If, on the contrary, the axis of the earth laid on the plane of the orbit, and remained ever parallel to itself, then the illuminated hemisphere would be bounded by a great circle, whose diameter would always be perpendicular to the earth's axis, and an equality of day and night would only occur when the earth held such a position that its axis would be perpendicular to the line joining the earth's center with the sun. Neither of these cases exists in nature, and, as we have already seen, the annual sweep of the sun from north to south, and from south to north, measures the double inclination of the earth's equator to the plane of the ecliptic, while the length of the day, as compared with the night, combined with the inclination of the solar beams, produces the alternation and changes of the seasons.

To an inhabitant of the earth's equator, the poles of the heavens will ever appear to lie in the horizon, and while the sun sweeps, during the year, from south to north, and returns, yet the days and nights are ever equal, and a perpetual summer reigns around the equatorial region, and a belt of extraordinary heat encircles the earth. Could an observer reach either pole of the earth, then the pole of the heavens would occupy his zenith, all diurnal circles would be parallel to the horizon, *which would now coincide with the equator*, and so long as the sun was south of the equator, (the observer being at the *north* pole of the earth), just so long would the sun be below the horizon, and every part of its diurnal circle would be invisible. On the day of the vernal equinox the sun would just reach the equator (now the horizon), and during the entire revolution would be seen sweeping round the horizon, slowly rising above it. This increase of elevation must now progress up to the summer solstice, and then decline to the autumnal equinox. The daylight thus continuing for six entire months, and the darkness for an equal length of time. These theoretic statements are abundantly verified by the facts, as reported by those who have visited high northern or southern latitudes.

Our climates are, then, undoubtedly, determined by the inclination of the earth's axis to the ecliptic, or what amounts to the same thing, by the inclination of the earth's equator to the ecliptic, the one angle being the complement of the other, or what it lacks of being ninety degrees.

The process employed by the ancients in measuring the inclination of the equator and ecliptic we have explained (chap. I.), and the same, with certain refinements, is still used by the moderns. At the beginning of the present century, this angle, called *the obliquity of the ecliptic*, amounted to 23° 27′ 56.″ 5. Two hundred and thirty years before Christ, the same angle, measured by the Greek astronomer, Eratosthenes, was 23.°51′.20.″ After a lapse of 370 years, Ptolemy found the inclination to be 23°.48′.45″. In the year 880 of our era, it was 23°.35′.00″. In 1690, Flamsteed found the same angle to be 23°.29′.00″, and thus from century to century the change progresses, reaching, however, a limit beyond which it cannot pass, (as we shall presently show), when it will commence a reverse motion, and thus the one plane slowly rocks to and fro upon the other in a calculable, but (so far as I know), not yet calculated period.

The time elapsing from the moment the earth is nearest the sun, until it returns again to the same point, is called an *annomalistic* year. The time from vernal equinox to the same again, is called a *tropical* year, while the time occupied by the earth in passing from any one point of its orbit, regarded as fixed, to the same point again, is called a *sidereal* year. These different periods, at the commencement of the current century, had the following values :—

Mean Annomalistic Year, in solar days,		. . .	365.2595981
" Tropical	"	" . . .	365.2422414
" Sidereal	"	" . . .	365,2563612

These figures being different, demonstrate the great and important fact that, whatever be the precise figure of the curve of the earth's orbit, the point of nearest approach to the sun, called the *perihelion*, is itself in motion. The same is true of the vernal equinox, the first evidently advancing, the second as evidently retrograding, and thus while the advance of the perihelion increases the length of the annomalistic year over the sidereal, the retrogression of the equinox decreases the length of the tropical, as compared with the sidereal year.

These figures are presented as the result of the best determinations which have been reached in modern times; but it must not be understood that the existence of these three different kinds of year are the discovery of our own times. The discovery of the motion of the vernal equinox, as we have seen, seems to reach back to the highest antiquity, and was known to all the ancient nations. The rate of motion was more exactly determined by the Greek astronomers, and hence the discovery has been attributed to that nation. Modern observations have confirmed this ancient discovery, while modern physical science has rendered a satisfactory account of this remarkable phenomenon, and has determined that the equinoctial point completes the entire circuit of the heavens in 25,868 years.

To ascertain the condition of the perihelion point as to rest or motion, it is only necessary to determine the sun's place among the fixed stars at the time of any perihelion, and to transmit the same to posterity. Any change of the sun's place among the stars at perihelion, which may become known in future ages, will demonstrate the fact that the perihelion is not only in motion, but will exhibit also the direction of the motion, and the rate of advance or recess. By a comparison of ancient observations with modern, the perihelion point of the earth's orbit is found to be slowly advancing, while, as we have stated above, the vernal equinox is slowly retrograding, at such rates that these two points pass each other once in 20,984 years. The perihelion coincided with the vernal equinox, as we are able to compute from their relative motions, 4,089 years before the Christian era. Sweeping onward to meet the summer solstice, the perihelion passed that point in the year twelve hundred and fifty of our era, and will meet the autumnal equinox about the year six thousand four hundred and eighty-three.

From the uniform rotation of the earth on its axis we obtain, as already stated, our unit of time. But this rotation is not sensible to man except by its effect on the position of objects external to the earth; and hence we determine the absolute period of rotation from marking, the moment when a fixed object, such as a star, passes the meridian of any given place. The time elapsing from this moment up to the next passage of the same object across the meridian, supposing the earth to be immovable as to its central point, would be the exact measure of the period of rotation of the earth on its axis. Now, the earth's center, in the space of one day and night, or during one rotation, actually passes over nearly 2,000,000 of miles, and it would seem as though this change of position would sensibly affect the return of our star to the meridian, but such is the vast distance of the fixed stars that visual rays sent to the same star, from the extremities of a base line of 2,000,000 miles in length, are absolutely parallel under the most searching instrumental scrutiny that man has been able to make. A sidereal day—the time which elapses between the consecutive returns of the same fixed star to any given meridian—is an invariable unit of time, and, as such, is extensively used in practical astronomy; but in civil life, inasmuch as all the duties of life are regulated by the return of the sun to the meridian, *solar*, and not *sidereal* time, has become the great standard in the record of all historic and chronologic events.' In case the earth did not revolve upon its axis, and had no motion except that of revolution in its orbit around the sun, it is manifest that in the course of one revolution the earth's axis, remaining parallel to itself, the circle dividing the illuminated from the dark hemisphere of earth would take up successively every possible position consistent with its always remaining perpendicular to the line joining the centers of the earth and sun. It is manifest, therefore, that by this revolution around the sun this luminary would be caused to rise above the horizon of any and every place upon the earth's surface successively, slowly to sweep across the heavens, and at the end of six months again to sink beneath the horizon. If, then, we define a solar day to be the time which elapses from the passage of the sun's center across any given meridian until it returns to the same meridian again, one such day would evidently be produced by the revolution of the earth in its orbit; hence we find a solar day to be longer than a sidereal day, because of the fact that the sun's center is brought to the meridian later, in consequence of its own apparent motion. Indeed, when we come to examine carefully the length of the solar day, we find it to be in a state of comparatively rapid change, a fact which we could readily have anticipated, as we know the apparent movement of the sun in its orbit, or rather the real motion of the earth, is changing from day to day. When the earth is in perihelion, or nearest the sun, it then travels with its greatest velocity, and passes over an arc of 1° 01′ 9″.9, in a mean solar day, whereas, when the earth is in aphelion, or furthest from the sun, it sweeps over an arc, in the same time, of only 57′ 11″.5. We thus perceive that the length of a true solar day must vary throughout the year, and for the purpose of obtaining a standard of time the world has adopted what is called a *mean* solar day, or a day having the *average* length of all the true solar days in the year. All the time-keepers employed in civil life, such as clocks and chronometers, are regulated to keep *mean solar time*, while, for the purposes of an observatory, *sidereal* time is in general use. This, however, is slightly different from the sidereal time already defined. The sidereal clock of the observatory, if perfectly true, would mark 0h. 00m. 00s. at the moment the *vernal equinox* is on the meridian of the observatory. It would mark the same at the next return, and hence this sidereal day is really a *vernal equinox day*. Now, as the sun's center appears to sweep round the whole heavens in the space of one year, and by virtue of this motion passes across the meridian of any place and returns to the same again, so, as we have seen, the vernal equinox sweeps around the heavens in a period of 25,868 years, and thus passes from one meridian to the same by virtue of this motion. Thus, a *vernal equinox day* is shorter than a sidereal day by an amount equal to one day in 25,868 years, a quantity very minute indeed, but still insisted upon, as we desire to impress upon the mind of the reader the differences between these various measures of time.

THE MOON A SATELLITE OF THE EARTH.—In prosecuting our plan of investigation we must now give some account of the moon, as she forms, astronomically speaking, a part of the planet which we call the earth, and we shall find hereafter that when we speak of the orbit, in which the earth revolves about the sun, the real point tracing that orbit is not the center of the earth, but a point determined by taking into consideration the fact that the earth and moon must be combined, as forming a sort of compound planet, revolving about the sun. Of all the celestial orbs furnishing objects for the investigation to man no one of them can rival the moon in the antiquity of its researches or in the importance and complexity of its revolutions.

If it were possible to trace the history of astronomical discovery, it would be found, beyond a doubt, that the first positive fact ever revealed to the student of the skies was *the motion* of the moon among the fixed stars. This fact is so obvious that any one who chooses to mark the moon's place by the stars which surround her to-night, and compare it with her place on to-morrow night, will make for himself the great discovery that the moon is sweeping around the heavens in a direction contrary to that of the diurnal revolution of the celestial sphere. Thus, if we mark the place of the new moon, in the evening twilight, when she appears as a silver crescent, emerging from the sun's beams, and just visible above the western horizon, we shall find that on the next evening, at the same hour, her distance from the horizon will have been greatly increased, and this increase of distance progresses from night to night, until we find the moon actually rising in the east at the time the sun is setting in the west. On the following night, at sunset, the moon will not have risen, but we will be compelled to wait nearly an hour after sunset before she becomes visible above the eastern horizon, and thus she advances in her orderly march among the fixed stars, until she circles entirely around the heavens, passes through the solar beams, and reappears in the west above the sun, as a slender crescent.

THE MOON'S REVOLUTION IN HER ORBIT.—We have already stated that, in case it were possible for the sun's center to trace out in its revolution among the fixed stars a line of golden light, visible to the eye of man, this line would be a regular circle, perfected at the close of one revolution, and ever after repeated, along the same identical track. Such, however, is not the case with our satellite. Could the moon's course be traced by leaving behind her among the stars a silver thread of light, at the completion of one revolution, this thread would not join on the point of beginning, but would be more or less remote, and the track described in the second and successive revolutions would not coincide with that first described; and thus we should find a multiplicity of silver lines sweeping round the circuit of the heavens, crossing each other, and interlacing in the most complicated manner, and thus making a girdle, or zone, of definite width, beyond whose limits the moon could never pass. The time occupied in completing one of these revolutions from a given star, until it returns to the great circle of the heavens, passing through the axis and this star again, is soon found to be variable, within certain narrow limits. This is called a *sidereal* revolution, and its mean value, at the beginning of the present century, is fixed at 27d. 7h. 43m. 11.5s. The most obvious lunar period, however, and that doubtless first discovered, is that called a *synodical* revolution, and is the period elapsing from the occurrence of full moon to full moon again, or from new moon to new moon again. The average length of this period, which is also called a *mean lunation*, amounted, at the epoch above mentioned, to 29d. 12h. 44m. 2s. 87. It is within the limits of this period that the moon passes through all those appearances which we call

THE MOON'S PHASES.—These extraordinary changes in the physical aspect of the moon must have perplexed the early astronomers. While the sun ever remained round and full-orbed in all his positions among the fixed stars, and while all the planets and bright stars shone with a nearly invariable light, the moon passed from a state of actual invisibility to a condition in which her disk was as round as that of the sun, and thence gradually losing her light, finally faded from the eye as she approached the solar orb. It was soon discovered that these changes were in some way dependent strictly upon the sun, and not upon the moon's place among the fixed stars. Any one who chooses may verify this discovery, for by locating the moon's place among the fixed stars at the full, and waiting her return to the same place again, it will be found that she has not yet reached her figure of a complete circle. Indeed, more than two days are required, after passing the position occupied when last full, before she gains the point that shall present us with a completely illumined disk. The discovery of this truth aided undoubtedly in solving the mystery of the moon's phases. It was clearly manifest that the moon was revolving about the earth in an orbit nearly circular. This was evident from the fact that the moon's apparent diameter did not change, by any sensible amount, during an entire revolution, which would have been impossible in case her approach to, or recess from the earth, had been very great in any part of her orbit.

Another phenomenon of startling interest aided greatly in reaching a true solution of the changes of the moon. I refer, of course, *to solar and lunar eclipses.* We have already referred to solar eclipses, as being undoubtedly produced by the interposition of the dark body of the moon between the eye of the spectator and the sun's disk. This demonstrated the fact that the moon in her revolution round the earth did sometimes cross the line joining the earth's centre with the sun, thus producing a central solar eclipse. It was thus manifestly possible for the moon's center to cross the same line at a point lying beyond the earth, with reference to the sun. When in this position, a straight line drawn through the center of the sun, and through the center of the earth, and produced onward, would pass through the moon's center, and to a person there situated, and looking at the sun, he would find the solar surface covered by the round disk of the earth, thus producing to the lunarian a solar eclipse. When the moon was thus situated, it was found to be shorn of a very large proportion of its light, not entirely fading from the eye, as did the sun when in total eclipse, but remaining indistinctly visible, with a dull reddish color. Now, as common observation teaches us that every opaque object casts a shadow in a direction opposite to the source of light, it follows that the earth must cast a shadow in a direction opposite to the sun; and in case this shadow reached as far as the moon's orbit, the moon, in taking up her successive positions, would sometimes pass into the earth's shadow. If self-luminous, the passage across the earth's shadow would occasion but a trifling change in her appearance. If, however, her light was either wholly or in greater part derived from the sun, then in passing into the earth's shadow, the stream of light from the sun being intercepted by the earth, the moon would lose her brilliancy, and could only be visible with an obscured lustre. All the phenomena presented in a solar as well as a lunar eclipse combine to demonstrate that the light of the moon is not inherent, or that this orb is not a self-luminous body; and all these phenomena were perfectly accounted for by admitting the hypothesis that the moon *shines* by *reflecting the light of the sun.* Thus, during a total solar eclipse, when the illuminated hemisphere of the moon was turned *from* the earth, her hither side appeared absolutely *black*, while no lunar eclipse ever occurred, except at a time when the moon's illuminated hemisphere was wholly visible, or at the full moon. In passing from new moon to full, it is evident, from the slightest reflection, that as the moon slowly recedes from the sun, in her movement round the earth, she will turn more and more of her illuminated hemisphere towards the earth, the whole of which will become visible when she is precisely opposite the sun, while the light must decrease in a reverse order in passing from the full moon to the new. Thus, all the facts and phenomena of ancient as well as of modern discovery combine to demonstrate the truth that the earth's satellite, like the planets already treated of, is only visible by reflecting the light of the sun.

We are ready by analogy to extend this reasoning to embrace the earth, and to believe that our own earth shines to the inhabitants of other planets (if such there be), by reflecting the light of the sun. We are not left, however, to mere analogy to demonstrate this truth, as we have the most positive evidence in the phases of the moon that the earth does reflect the solar light. No one can have failed to notice the fact that when the moon appears as a slender crescent, her *entire disk* may be traced, faintly visible even to the naked eye; but when the telescope is applied, we readily distinguish in this darkened part all the outlines and prominent features which become visible to the unaided eye when the moon is entirely full. This faint luminosity is beyond all doubt occasioned by the reflection back again to the earth of that light which the earth reflects upon the moon; for if we consider the relative positions of the sun, moon, and earth, we shall see that at the new moon the whole illuminated hemisphere of the earth is turned full upon her satellite, and at that time the largest amount of light from the earth falls upon the surface of the moon. The relative positions of the bodies now slowly change, and as the moon increases in light by like degrees, the earth loses in light; and when the moon becomes entirely full, the earth will be to the lunarian entirely dark, as her non-luminous hemisphere is then turned directly to the moon.

We have already stated that, during a lunar eclipse, the moon remains dimly visible. This is not due to the reflected light of the sun, thrown upon the moon by the earth, but arises from the fact that the solar rays are so much bent out of their course in passing through the earth's atmosphere, that many of them are still able to reach the moon's surface, and thus in some degree to light up her disk, even during a central eclipse.

Amid all the variations and changes which mark the luminosity of the moon one thing remains almost absolutely invariable. No eye on earth has yet seen more than one half of the lunar sphere. The hemisphere now visible to man, has (so far as we know,) ever been visible, and, except by the intrusion of some foreign body, will ever remain turned toward the earth. There are slight deviations from the positiveness of this statement to which we shall have occasion to allude hereafter, but the grand truth remains, that the same hemisphere of the moon is ever turned toward the earth.

To account for this remarkable fact we are compelled to acknowledge a rotation of the moon on her axis, in the exact period employed by her in her revolution in her orbit. If the moon had no motion of rotation about an axis, then in the course of her orbital revolution every portion of her surface would come into view successively.

This explanation, which it would seem ought to be perfectly satisfactory, has, in some strange way, been not only misunderstood, but denied; and yet should the person most skeptical undertake to walk round a central object, always turning his face to the center, without as well turning his shoulders and person, he would receive a positive conviction of the truth of our explanation of a most practical character.

The physical cause of this remarkable fact in the moon's history will be duly considered hereafter.

The same kind of observation and reasoning which enabled Hipparchus to determine the eccentricity of the sun's apparent orbit (the earth's real orbit) sufficed to enable this philosopher to determine the eccentricity of the moon's orbit, and the epicyclical theory gave a tolerably fair account of the most striking irregularities in the moon's motion. In one respect, however, we find a remarkable difference between the lunar and solar motions. The position of the perihelion of the earth's orbit moves so slowly that for a period of even a hundred years this motion may be neglected without any great error. While the moon's *apogee*, or least distance from the earth, was found to be sweeping round the heavens with a comparatively rapid motion, following the moon in her course among the stars, so that while in a period of 6,585½ days the moon performed 241 complete revolutions with reference to the stars, she made but 239 revolutions with regard to her perigee. Hipparchus succeeded in representing this motion by means of eccentrics and epicycles, and finally was able to tabulate the moon's places with such accuracy as to represent her positions, especially at the new and the full, so as to predict roughly solar and lunar eclipses.

Ptolemy discovered, 500 years later, a new irregularity in the moon's motion, which reached its maximum value in what are called the *octants*, that is, the points halfway between the new moon and her first quarter, and so on a quarter of a circumference in advance round the orbit. New attempts were made to explain these irregularities by a combination of circles and eccentrics. It was, finally, approximately accomplished, but all these facts thus accumulating were preparing the way for the abandonment of an hypothesis which could only be maintained by the imperfection of astronomical observation.

The excursions made by the moon, north and south of the ecliptic, or plane of the earth's orbit, were obviously to be accounted for by the fact that this satellite revolved in a plane, inclined under a certain angle, to the ecliptic. This angle was readily measured by the ancients, and, though slightly variable, was fixed at the beginning of our century at 5° 8′ 47″.9.

THE LUNAR PARALLAX AND DISTANCE.—The rude instruments employed by the early observers in their astronomical observations were insufficient for any delicate work, and hence we find them quite ignorant of the absolute value of even the moon's parallax, a quantity which far exceeds any other parallactic angle of the solar system. We have already shown (Chap. I.) how the distance of an inaccessible object may be obtained by measuring the angles formed at the extremities of a given base line, by visual rays drawn to the object. In case the base line be very short in proportion to the distance to be measured, the sum of the two angles thus measured will approach in value 180°, and the angle at the distant object formed by the visual rays becomes smaller in proportion to its distance. In our attempts to measure the solar parallax, using the earth's diameter as a base, it was found that the delicacy of modern instruments was not adequate to so difficult a task. This, however, is not the case when we come to apply them in the determination of the lunar parallax. Indeed, the moon is found to be so near the earth that visual rays, drawn from spectators at different parts of the earth, not very remote from each other, to the moon's center, form with each other sensible angles; and thus the moon, viewed from different stations, is projected among different stars. When the moon's center is in the absolute horizon, (that is, in a plane passing through the center of the earth and perpendicular to the earth's radius drawn to the place of the spectator), lines drawn from the center of the

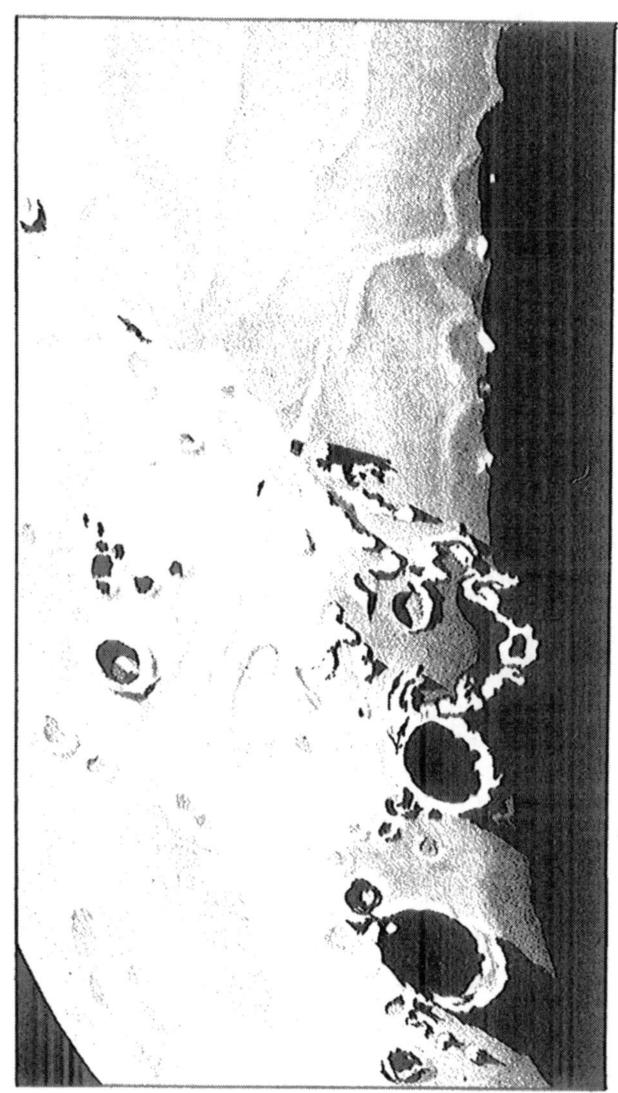

NORTH WESTERN BOUNDARY OF MARE SERENITATIS
1860 FEBR. 27 8 H. P.M. ALBANY TIME
DUDLEY OBSERVATORY.

earth and from the eye of the observer unite at the moon's center, under an angle called the moon's *horizontal* parallax. In case the moon's distance from the earth were constant, this angle would also be invariable. This, however, is not the case, and we find the horizontal parallax reaches a maximum value equal to 1° 1′ 24″, when the moon is nearest the earth, and a minimum value of 0° 53′ 48,″ when most remote—the average value being 0° 57′ 00″.9. These angles give for the moon's mean distance from the earth 237,000 miles.

As all the computed places of the planetary orbs assume the spectator to occupy the earth's center, we readily perceive that, in the case of the moon, the computed and observed places would never agree, except in one instance, namely, that in which a line joining the center of the earth with the moon's center passes through the place of the observer, or when the moon's center is exactly in the zenith. The effect of parallax on the apparent place of the moon is to sink it below the position it would have held in case it were seen from the earth's center.

Knowing the actual distance of the moon, her real diameter is readily determined, and is found to be about 2,160 miles; hence her volume is about one-forty-ninth part of that of the earth. We shall have occasion hereafter to resume our examination of the moon's motions when we come to discuss the physical causes by whose power the planetary orbs are held in dynamical equilibrium, and are retained in their orbits. We now proceed to examine the physical constitution of

THE MOON, AS REVEALED BY THE TELESCOPE.—The splendid instruments which modern skill and science have furnished for the examination of the distant worlds so far increase the power and reach of human vision, in the case of the moon, as to bring this satellite of the earth comparatively within our reach. A telescope which bears a magnifying power of one thousand times, applied to the examination of the moon's surface, enables the observer to approach to within 237 miles of this extraordinary world, and even this distance, under the most favorable circumstances, may be reduced by one-half. This, perhaps, is the nearest approach ever made to the moon, and it is at a distance of say 150 miles that we are permitted to stand and examine at our leisure the features which diversify the surface of our satellite. No subject has excited so deep an interest from mere curiosity, as that involved in the actual condition of the moon's surface. Every one desires to know if the other worlds are like our own. Have they oceans and seas, lakes, rivers, islands, and continents? Does their soil resemble our own? Does vegetable life there manifest itself in every variety of grass and flowers, and shrub and tree? Are there extended forests and spicy groves, filled with multitudinous animals, in these far off worlds? And, above all, are these bright orbs inhabited by rational intelligent beings like man? The earnest desire to obtain responses to these and like questions, caused to be received, many years since, with the most wonderful delight and credulity, a statement put forth in America, giving professedly the details of lunar discoveries, said to have been made by Sir John Herschel at the Cape of Good Hope, in which all these questions were most satisfactorily answered. We need hardly say how great was the disappointment when these pretended discoveries proved to be but fanciful inventions. When we call to mind that with a telescope magnifying 2,000 times we are still separated from the moon 120 miles, we readily perceive the utter impossibility of solving at present, directly by vision, the problem of the moon's habitability. We know not what may be accomplished by human genius and human invention, and after the production of so marvellous an instrument as a telescope capable of transporting the beholder to within 120 miles of the surface of a body actually removed 237,000 miles, we will not presume to set any specific limits to future effort. We can only say that the telescope must become vastly improved in its powers of definition and development before we can hope to satisfy ourselves, from actual inspection, that our satellite is or is not inhabited by a race with any of the faculties which distinguish man.

Let us see what has actually been accomplished by telescopic investigation, and although it falls far short of satisfying the curiosity of our nature we shall find much to interest and astonish. We can affirm, then, that the surface of our satellite is diversified with hill and dale, with lofty mountains and mighty cavities, with extensive plains and isolated mountain peaks, not very unlike the same features presented by our earth. The hemisphere of the moon, visible to man, has been studied and mapped with the greatest care. Indeed, its elevations and depressions have been accurately modeled, the mountain elevations have been measured, and the depths of the mighty cavities which distinguish her surface have all been carefully determined. These measures all depend on the fact that the moon receives its light from the sun, and presents its surface to that orb under every angle in the course of its revolution. The mountains of the moon, like those of the earth, have their summits first lighted by the rays of the rising sun, while all the plain beneath, and their rough and rugged sides, are in the deepest darkness. These summits, when so illuminated, glow and sparkle with a dazzling beauty unsurpassed. As the sun rises, we perceive distinctly the black shadow of the mountain falling to a great distance on the plain below. These shadows slowly decrease in length, and their outlines gradually creep up the mountain side as the sun reaches the moon's meridian. When the sun begins to decline the shadows fall in the opposite direction, slowly extend their black masses over the distant plains, and darkness finally gathers round the mountain sides, till again the summit is alone illumined by the rays of a setting sun. It is by means of those shadows, whose lengths are readily determined by micrometrical measures, that we are enabled to determine the heights of the lunar mountains and the depths of the lunar cavities. This process is not more difficult than to determine the elevation of a church steeple or other lofty object by the length of its shadow cast upon a horizontal plane below. The altitude of the sun above the horizon at noon will give the direction of the visual ray passing from the summit of the object to the extremity of its shadow. Knowing the value of this angle, and the measured length of the shadow cast, we have at once the means of determining the elevation of the object under examination. These simple principles are readily transferred to the determination of the heights and depths of the lunar surface, while the figure of the shadow cast by the summits of a mountain range on an extended plain below, gives to us almost as perfect a knowledge of the actual forms of the lunar mountains as though it were possible actually to tread their lofty summits.

We find upon the moon's surface a range of mountains lifting themselves above a level country and extending

LUNAR SURFACE

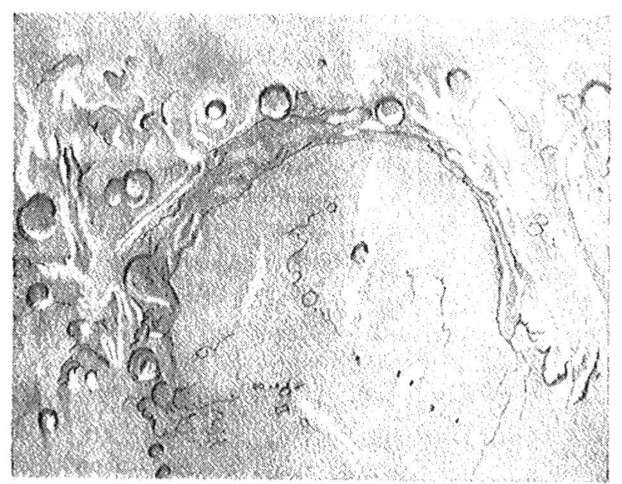

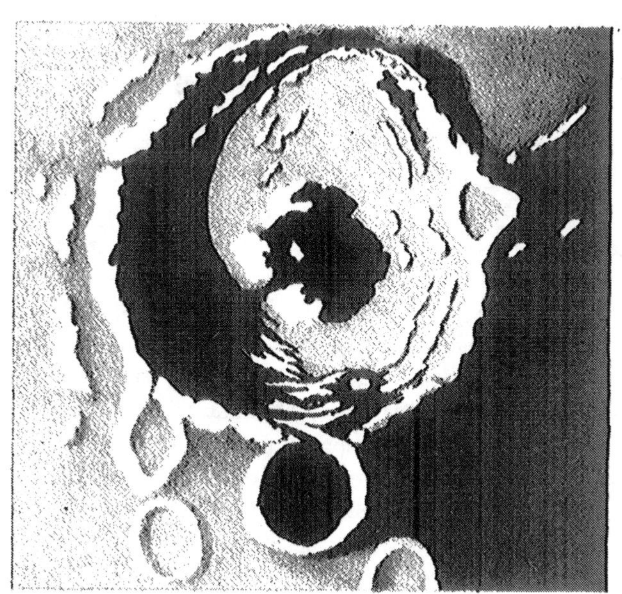

GASSENDIUS, DUDLEY OBSERVATORY
JAN. 1860.

nearly two hundred miles, which have received the name of the *Appenines*. This mountain range comes into the sunlight just after the moon has passed its first quarter, and is then one of the finest objects that the telescope reveals to the eye of man. The brilliancy of the illuminated heights and ridges, the absolute blackness of the deep, rocky chasms, the lofty peaks, the rugged precipices, and the deep shadows, all combine to increase the natural grandeur of this extensive mountain range. Let it not be imagined that details in such a scene, such as actual individual rocks, of definite form and outline, are to be seen; but as lights and shades produce the forms of every surface, so these lights and shadows on the moon bring out the absolute forms in the most distinct and perfect manner. The contrasts between the dark and illuminated parts of the moon are far deeper and stronger than on the earth. This arises from the fact that the sunlight on the moon is not reflected or refracted by an atmosphere such as surrounds the earth. The twilight which attends the setting sun and the dawn, which so beautifully announces the coming of day, does not exist for the lunarians. If any eye beholds the rising of the mighty orb of day from those lofty lunar summits which are first illumined by his horizontal beams, no gentle flashings, or rosy tints, or purple hues, but from intense darkness there is an instantaneous burst of brilliant sunlight. The beauty of our dawns and twilights is due to the atmosphere which surrounds the earth, and while we cannot affirm that no such atmosphere surrounds our satellite, we are certain that whatever gaseous envelope may surround the moon on its hither side, its density cannot compare with that of the terrestrial atmosphere. Under very favorable circumstances, with the great refractor of the

Cincinnati Observatory, the author has either seen, or fancied he saw, a faint penumbra edging the dark mountain shadows, and clinging to the black outline, as it slowly crept up the mountain side, as the sun rose higher and higher. We shall return to this subject when we come to treat of certain peculiarities attending the eclipse of the sun, and the occultation of stars by the moon.

Some of the mountains of the moon reach an elevation of 8 to 10,000 feet above the general level. Here and there we find insulated peaks rising abruptly from extended plains to a height of 6 or 7,000 feet, and in the early lunar morning flinging their long, sharp, black shadows to a vast distance.

But the most remarkable feature presented in the lunar surface is the tremendous depths of some of the cavities, and their immense magnitude. Some of them extend beneath the general level of the country to a depth of 10 to 17,000 feet, and their rough, misshapen, precipitous sides, exhibit scenes of rugged sublimity to which earth presents no parallel. Of these cup-shaped cavities, especially in the southern portion of the lunar hemisphere, the number is beyond credibility; and, in case we admit them to be the extinct craters of once active volcanoes, we are forced to the conclusion that convulsions, such as the earth is a stranger to, have shaken the outer crust of our satellite into a hideousness of form unknown in any region of our planet. Some of these deep cavities are nearly circular in figure, and with diameters of all magnitudes up to twenty miles. Very often the interior will exhibit a uniformly shaded surface, and in the center a conical mountain will lift itself far above this level plain. That these convulsions are of different ages is clearly manifest from the fact that their outlines very often overlap one another, and the oldest and the newest formations are thus distinctly traced by the eye of man. So sharp and positive is the outline of these extraordinary objects that one cannot but feel that some sudden bursting forth might even occur while under telescopic examination. Once indeed, while closely inspecting these seemingly volcanic mountains and craters of the moon, I was startled by a spectacle which, for a moment, produced upon the mind a most strange sensation. A mighty bird, huge in outline and vast in its proportions, suddenly lifted itself above the moon's horizon and slowly ascended in its flight towards the moon's center. It was no lunar bird, however, but one of earth, high up in the heavens, winging its solitary flight in the dead of night, and by chance crossing the field of vision and the lunar disk.

Before the power of the telescope had reached its present condition of perfection the darker spots of the moon were assumed to be seas and oceans; but the power now applied to the moon demonstrates that there cannot exist at this time any considerable body of water on the hemisphere visible from the earth. And yet we find objects such, that in case we were gazing upon the earth from the moon, possessing our actual knowledge of the earth's lakes and rivers, we should pronounce them, without hesitation, lakes and rivers. There is one such object which I will describe as often seen through the Cincinnati Refractor. The outline is nearly circular, with a lofty range of hills on the western and south-western sides. This range gradually sinks in the east, and a beautiful sloping beach seems to extend down to the level surface of the inclosed lake (as we shall call it, for want of other language). With the highest telescopic power, under the most favorable circumstances, I never could detect the slightest irregularity in the shading of the surface of the lake. Had the cavity been filled with quick-silver and suddenly congealed or covered with solid ice, with a covering of pure snow, the shading could not be more regular than it is. To add, however, to the terrene likeness, into this seeming lake there flows what looks exactly as a river should at such a distance. That there is an indentation in the surface, exactly like the bed of a river, extending into the country, (with numerous islands,) for more than a hundred miles, and then forking and separating into two distinct branches, each of which pursues a serpentine course for from thirty to fifty miles beyond the fork, all this is distinctly visible. I may say, indeed, that just before entering the lunar lake this lunar river is found to disappear from sight, and seems to pass beneath the range of hills which border the lake. The region of country which lies between the forks or branches of this seeming river, is evidently higher, and to the eye appears just as it should do, so as to shed its water into the stream which appears to flow in the valley below. The question may be asked, why is this not a lake and a river? There is no lunar atmosphere on the visible hemisphere of the moon, such as surrounds the earth, and if there were water like ours on the moon, it would be soon evaporated, and would produce a kind of vaporous atmosphere, which ought to be shown in some of the many phenomena involving the moon, but has not yet been detected. What, then, shall we call the objects described? I can only answer that this phenomenon, with many other, presented by the lunar surface, has thus far baffled the most diligent and persevering efforts to explain. In some of these cavities, where the tinting of the level surface is so perfect with an ordinary telescope, when examined with instruments of the highest power, we detect small depressions in this very surface, cup-shaped, and in all respects resembling the form and features of the principal cavity. These hollow places are clearly marked by the shadows cast on the interior of the edges, which change as the sun changes, and seem to demonstrate that these level surfaces do not belong to a fluid but to a solid substance.

Among what are called the volcanic mountains of the moon are found objects of special interest. One of them, named Copernicus, and situated not far from the moon's equator, is so distinctly shown by the telescope, that the external surface of the surrounding mountains presents the very appearance we would expect to find, in mountains formed by the ejecting from the crater, of immense quantities of lava and melted matter, solidifying as it poured down the mountain side, and marking the entire external surface with short ridges and deep gullies, all radiating from a common center. Can these be, indeed, the overflowing of once active volcanoes? Sir William Herschel once entertained the opinion that they were, and, with his great reflecting telescope, at one time discovered what he believed to be the flames of an active volcano on the dark part of the new moon. More powerful instruments have not confirmed this discovery, and although a like appearance of a sort of luminous or brilliant spot, has been seen by more than one person, it is almost impossible to assert the luminosity to be due to a volcano in a state of irruption, but is more commonly supposed to be some highly reflective surface of short extent, and for a time favorably situated to throw back to us the earth-shine of our own planet.

From some of these seeming volcanoes there are streaky

radiations or bright lines, running from a common center, and extending sometimes to great distances. These have by some been considered to be hardened lava streams of great reflective power, but, unfortunately for this hypothesis, they hold their way unbroken across deep valleys and abrupt depressions, which no molten matter flowing as lava does, could possibly do. To me they more resemble immense upheavals, forming elevated ridges of a reflecting power greater than that of the surrounding country.

We find on the level surfaces a few very direct *cuts*, as they may be called, not unlike those made on our planet for railway tracks, only on a gigantic scale, being more than a thousand yards in width, and extending in some instances over a hundred miles in length. What these may be it is useless to conjecture. We cannot regard them as the work of sentient beings, and must rather consider them as abrupt depressions or faults in the lunar geography.

THE MOON'S CENTER OF FIGURE.—The wonderful phenomena presented to the eye on the visible hemisphere of the moon have been rendered in some degree explicable by a remarkable discovery recently made, that the *center* of *gravity* of the moon does not coincide with the *center* of *figure*. This is not the place to explain *how* this fact has been ascertained. It is now introduced to present its effect, on the hither portion of the lunar orb.

If the material composing the moon was lighter in one hemisphere than the other, it is manifest that the center of gravity would fall in the heavier half of the globe. For instance, a globe composed partly of lead and partly of wood could not have the center of gravity coincident with the center of the globe; but it would lie somewhere in the leaden hemisphere. So it now appears that the center of gravity of the moon is more than 33 miles from the center of figure, and that this center of gravity falls in the remote hemisphere, which can never be seen by mortal eye.

Now, the center of gravity, is the center to which all heavy bodies gravitate. About it as a center the lunar ocean and the lunar atmosphere, *in case such exist, would arrange themselves*, and the lighter hemisphere would rise above the general level, as referred to the center of gravity, to an extreme height of 33 miles. Admitting this to be true, and as we shall see hereafter the fact appears to be well established, we can readily perceive that no water, river, lake or sea, should exist on the hither side of the moon, and no perceptible atmosphere can exist at so great an elevation. Even vegetable life itself could not be maintained on a mountain towering up to the enormous height of 33 miles; and hence we ought to expect the hither side of our satellite to present exactly such an appearance as is revealed by telescopic inspection.

If the centers of gravity and figure ever coincided in the moon, and the change of form has been produced by some great convulsion, which has principally expended its force in an upheaval of the hither side of the globe, then we can account for the rough, broken, and shattered condition of the visible surface. Lakes and rivers may once have existed, active volcanoes might once have poured forth their lava streams, while now the dry and desolate beds and the extinct craters are only to be seen.

The consequences which flow from this singular discovery as to the figure of our satellite are certainly very remarkable, and will doubtless be traced with deep interest in future examinations.

OCCULTATIONS.—As the moon is very near the earth, and her disk covers a very considerable surface in the heavens in her sweep among the fixed stars, she must of course cross over a multitude of stars in her revolutions. A star thus hidden by the moon is said to be *occulted*, and these *occultations* are phenomena of special interest on many accounts. As a general thing, a star even of the first magnitude, in passing under the dark limb of the moon, vanishes from the sight instantaneously, as though it were suddenly stricken from existence, and at its reappearance its full brilliancy bursts at once on the eye. This demonstrates the fact that the stars can be nothing more than luminous points to our senses, even when grasped by the greatest telescopic power.

A strange appearance sometimes attends the occultation of stars by the moon. The star comes up to the moon's limb, entirely vanishes for a moment, then reappears, glides on the bright limb of the moon for a second or more, and then suddenly fades from the sight.

This phenomenon, as also another of most startling character attending sometimes the total eclipse of the sun, when blood-red streaks in radiations are found to shoot suddenly from behind the moon's limb, are supposed by some to demonstrate the existence of a lunar atmosphere. Much attention has been bestowed on the total eclipses of the sun during the past twenty years, for the express purpose of solving, if possible, these mysterious radiations of red light. Some entertain the opinion that they are due to the colored glasses used to soften the intense solar light, as seen through the telescope. We can only say that these phenomena remain without satisfactory explanation, and that the physical condition of the moon is yet a problem of the deepest interest. We can assert the irregularities of her surface, her deep cavities and lofty elevations, her extended plains and abrupt mountain peaks, but beyond this our positive knowledge does not extend.

We shall resume the consideration of our satellite when we come to discuss the great theory of universal gravitation.

CHAPTER V.

MARS, THE FOURTH PLANET IN THE ORDER OF DISTANCE FROM THE SUN.

PHENOMENA OF MARS DIFFICULT TO EXPLAIN WITH THE EARTH AS THE CENTER OF MOTION.—COPERNICAN SYSTEM APPLIED.—EPICYCLE OF MARS.—BETTER INSTRUMENTS AND MORE ACCURATE OBSERVATIONS.—TYCHO AND KEPLER.—KEPLER'S METHOD OF INVESTIGATION.—CIRCLES AND EPICYCLES EXHAUSTED. —THE ELLIPSE.—ITS PROPERTIES.—THE ORBIT OF MARS AN ELLIPSE.—KEPLER'S LAWS.—ELLIPTICAL ORBITS OF THE PLANETS.—THE ELEMENTS OF THE PLANETARY ORBITS EXPLAINED.—HOW THESE ELEMENTS ARE OBTAINED. —KEPLER'S THIRD LAW.—VALUE OF THIS LAW.—THE PHYSICAL ASPECT OF MARS.—SNOW ZONES.—ROTATION OF THE PLANET.—DIAMETER AND VOLUME.—SPECULATION AS TO ITS CLIMATE AND COLOR.

THIS planet is distinguished to the naked eye by its brilliant red light, and is one of the planets discovered by the ancients. To the old astronomers Mars presented an object of special difficulty. Revolving as it does in an orbit of great eccentricity, sometimes receding from the earth to a vast distance, then approaching so near as to rival in brilliancy the large planets, Jupiter and Venus, on the old hypothesis of the central position of the earth, and the uniform circular motion of the planets, Mars presented anomalies in his revolution most difficult of explanation.

These complications were measurably removed by the great discovery of Copernicus, which released the earth from its false position, and gave to Mars its true center, the sun; but even with this extraordinary advance in the direction towards a full solution of the mysterious movements of this planet, there remained many anomalies of motion of a most curious and incomprehensible character. It will be remembered that Copernicus, in adopting the sun as the center of the planetary orbits, was compelled to retain the epicycle of the old Greek theorists, to account for the facts which still distinguished the planetary revolutions. As in the revolution of the earth about the

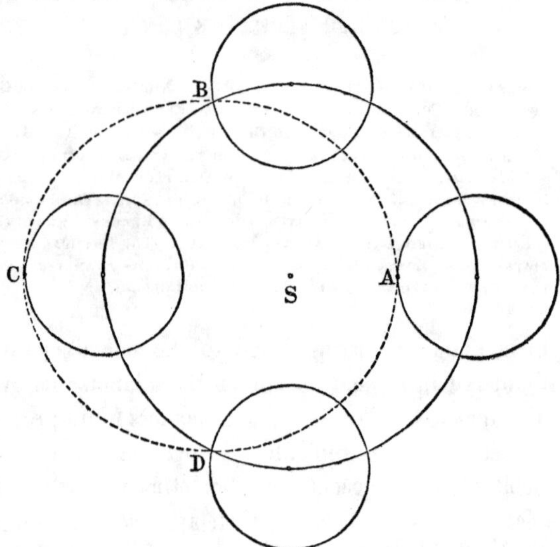

sun there was an approach to and recess from this central orb, so in the revolution of Mars it was manifest that there was a vast difference between the aphelion and perihelion distances of the planet. The epicycle was then retained to account for this anomaly in the motion of Mars; and it will be readily seen from the figure above how this hypothesis rendered a general explanation of the facts presented for examination.

The large circle, having the sun for its center, represents the orbit of Mars, that is, a circle whose radius is equal to the average or mean distance of the planet. The small circles represent the epicycle, in the circumference of which the planet revolves with an equable motion, while *its center* moves uniformly round on the circumference of the large circle. When the planet is at A, it is in perihelion, or nearest the sun. While the center of the epicycle performs a quarter revolution, the planet also performs in its epicycle a quarter of a revolution, and reaches the position B. A half revolution brings it to *aphelion* in C, and three quarters of a revolution in the epicycle locates the planet at D, and an entire revolution brings it again to A, the point of departure. Thus it will be seen that the planet must describe an oval curve, traced in the figure A B C D, and for general purposes this exposition of the phenomena seemed entirely satisfactory. It is true that it only accounted for the movement from east to west, or in longitude, while the motion north and south of the earth's orbit, or in latitude, was accounted for by supposing the plane of the epicycle to vibrate or rock up and down, or right and left of the plane of the ecliptic, while its center moved uniformly round in the great circle constituting the orbit of the planet.

So long as observation was so defective as to yield but rough places of the heavenly bodies, the deviations from the path marked out by the theory of epicycles escaped detection. The erection of the great observatory of Uraniberg, by the celebrated astronomer Tycho Brahe, and the furnishing it with instruments of superior delicacy, introduced a new era in the history of astronomical observation. The instruments employed by Copernicus were incapable of giving the place of a star or planet with a precision such as to avoid errors amounting to even the half of one degree, or an amount of space equal to the sun's apparent diameter. The instruments employed by Tycho reduced the errors of observation from fractions of degrees to fractions of minutes of arc, and when thus critically examined, the planets, as well as the sun and moon, presented anomalies of motion, requiring to account for them a large accumulation of complexity in the celestial machinery. Such was the condition of theoretic and practical astronomy at the era inaugurated by the appearance of the celebrated Kepler. This distinguished astronomer early became a devoted advocate of the Copernican system of the universe, adopting not only the central position of the sun, but also the ancient doctrine of uniform circular motion, and the theory of epicycles. The investigations of Kepler on the motions of the planet Mars commenced after joining Tycho at Uraniberg, in 1603, and, based upon the accurate observations of this later astronomer, finally led to the overthrow of the old theory of epicycles and circular motion, introduced the true figure of the planetary orbits, and with the elliptical theory of planetary motion, commenced the dawn of that brighter day of modern science, which in our age sheds its light upon the world.

The history of the great discoveries of Kepler presents one of the most extraordinary chapters in the science of astronomy. It must be remembered that the doctrine of circular motion, at once so beautiful and simple, had held its sway over the human mind for more than two thousand years. Such, indeed, was its power of fascination that even the bold and independent mind of Copernicus could not break away from its sway. When Kepler commenced his examination of the movements of Mars it was under the full and firm conviction that the theory of circles and epicycles was unquestionably true. His task, then, was simply to frame a combination such as would account for the new anomalies in the motions of Mars discovered by the refined observations of Tycho. The amount of industry, perseverance, sagacity, and inventive genius displayed by Kepler in this great effort is unparalleled in the history of astronomical discovery. His plan of operation was admirably laid, and if fully and faithfully carried out, could not fail, in the end, to exhaust the subject, and to prove at least the great negative truth, that no combination of circles and epicycles could by any possibility truly represent the exact movements of this flying world. It is useless to enumerate the different hypotheses employed by Kepler. They were no less than nineteen in number, each of which was examined with the most laborious care, and each of which, in succession, he was compelled to reject. Having adopted an hypothesis, he computed what ought to be the visible positions of the planet Mars, as seen from the earth, throughout its entire revolution. He compared these *computed* places or positions with the *observed* places, or those actually occupied by the planet, and finding a discrepancy between the two, his hypothesis was thus shown to be false and defective, and must necessarily be rejected.

It is curious to note the limits of accuracy in the observed places of the planet, upon which Kepler relied with so much confidence in this bold investigation. Many of the various hypotheses which he worked up and applied with so much diligence, enabled him to follow the planet in its entire revolution around the sun, with discrepancies between *observation* and *computation* not exceeding the tenth part of the moon's diameter. Indeed, the whole error in the computed place of Mars, when compared with its observed place, when Kepler commenced the problem, did not exceed *eight minutes* of arc, or about one-fourth of the moon's apparent diameter, and yet upon this slender basis this wonderful man declared that he would reconstruct the entire science of the heavens.

Having thus framed one hypothesis after another, each of which was in its turn rigorously computed, applied and rejected, this exhaustive process finally brought Kepler to the conclusion that no combination of circles, with circular motion, could render a satisfactory account of the anomalies presented in the revolution of Mars; and he thus rose to the grand truth, that the circle, with all its beauty, simplicity, and fascination, must be banished from the heavens.

The demonstration of this great negative truth was a necessary preliminary to the discovery of the true orbit in which Mars performed his revolution around the sun. Complexity having been exhausted in the combination of circles without success, Kepler determined to return to primitive simplicity and endeavor to find some *one* curve which might prove to be that described by the planet. In tracing up the movement of Mars, as we have seen, the figure of the true orbit was evidently an *oval*, and among ovals there is a curve known to geometricians by the name of the *ellipse*. This curve is symmetrical in form, and enjoys some peculiar properties which we will exhibit to the eye.

The line A B is called the *major axis*, and is the longest line which can be drawn inside the curve. It passes from one *vertex* A to the other *vertex* at B, and the semi-ellipse A D B is such that if turned round the axis A B, it would fall on, and exactly coincide with the semi-ellipse, A C B. The line C D is called the *minor axis*, and is the *shortest line* which can be drawn in the ellipse. This line divides the figure into two equal portions, exactly symmetrical.

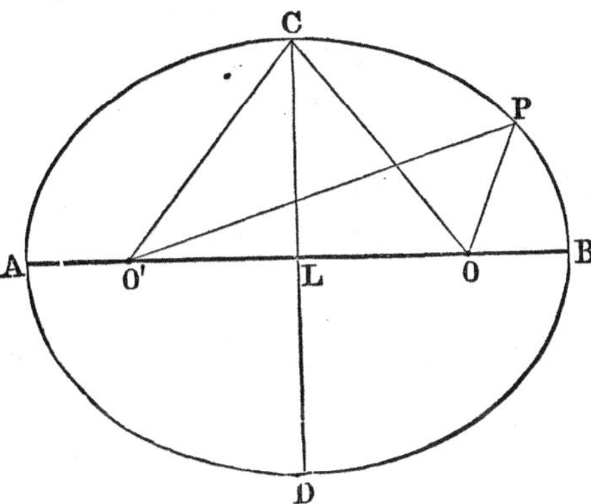

The point L is called the *center* of the ellipse, and divides all the lines drawn through it and terminating in the curve into two equal parts. But there are two points, O and O′, called the *foci*, which enjoy very peculiar properties. If from C as a center, and with a radius equal to A L, the semi-major axis, we describe an arc, it will cut the major axis in O and O′, the two foci. Now, in case we assume any point on the curve as P, and join it with O and O′, the sum of these lines, O P and O′ P, will be equal to the major axis, A B.

Such are the distinguishing properties of the curve, which holds the next rank in order of beauty, simplicity, and regularity, after the circle. While the circle has one central point, from which all lines drawn to the curve are equal, the ellipse has *two foci*, from which lines drawn to the same point on the curve, when added together, are equal in length to the major axis. When the major axis of the ellipse is assumed as the diameter of a circle, the circumference will wholly inclose the ellipse. When the minor axis is assumed as the diameter, the circumference will lie wholly within the ellipse. When the foci, O and O′, are very near the center, then these circles, and the ellipse lying between them, are very close to each other.

When Kepler was compelled to abandon the circle and circular motion as a means of representing the planetary revolutions, he adopted the ellipse as the probable form of the orbits of these revolving worlds, and made an especial effort to apply this new figure to a solution of the mysteries which still enveloped the motions of Mars. But here a new difficulty presented itself. In the circular orbits and epicycles a uniform motion was always accepted, but in the ellipse, every point of which is at unequal distances from the focus, some law of velocity had to be discovered to render it possible to compute the planet's place, even after the axis of the ellipse had been determined. Here again was opened up to the mind of the laborious philosopher a wide field of investigation. Many were the hypotheses which he framed, computed, applied and rejected, but finally fixing the sun in the focus of the assumed elliptic orbit, and assuming that the line drawn from the sun's center to the planet would sweep over *equal amounts* of *area in equal times*, he computed the places of Mars through an entire revolution. These newly computed places were now compared with those actually filled by the revolving world, and Kepler found to his infinite delight that the planet swept over the precise track which his hypothesis had enabled him to predict, and with an exultation of victorious triumph to which the history of pure thought furnishes few parallels, Kepler announced to the world his *two first laws* of planetary motion, which may be given as follows: —

1. *Every planet revolves in an elliptical orbit about the sun, which occupies the focus.*

2. *The velocity of the planet on every point of its orbit is such that the line drawn from the sun to the planet will sweep over equal areas in equal times.*

At the time Kepler lived, human genius could not have won a grander triumph, for it was not only a triumph over nature, which compelled her to render up her inscrutable secrets, but a triumph which for ever freed the mind from the iron sway of the schools, and from the prejudices which had become venerable with the lapse of more than twenty centuries of unyielding power. No grander emotions ever swelled the human heart than those which Kepler experienced when, tracing this fiery world through his sweep among the fixed stars, he found he had truly and firmly bound his now captive planet in chains of adamant, from which in all future ages it could never escape, having fixed for all time the figure of the orbit and the law of its orbital velocity. This extended notice is due to the well merited fame of Kepler, as well as to the grandeur of the laws discovered.

The elliptical theory, now successfully applied to the planet Mars, was extended rapidly to Mercury, to the moon, and in order to all the known planets. We shall hereafter, in our treatment of the planets, adopt the elliptical theory, and to render our language entirely intelligible, will proceed to explain what is meant by the *elements* of the orbit of a planet.

To determine the *magnitude* of any ellipse, we must know the longer and shorter axis, or the longer axis and the distance from the center to the focus, called the *eccentricity*.

To determine the position of the plane of an ellipse, we must know the position of the line of its intersection with a given plane, (usually the ecliptic) called the *line of nodes*, and also the *angle of inclination* with this fixed plane.

To determine the *position* of the elliptical orbit in its own plane, we must know the position of the *vertex*, or extremity of the major axis, called the *perihelion*.

And finally, to trace the planet after all these matters shall be known as to its orbit, we must know *its place* or position in its orbit at a given moment of time, and its period of revolution.

Now, every plane of every planetary orbit passes through the sun's center.

Every longer axis of every planetary orbit passes through the sun's center, and every *line* of *nodes* of all the planetary orbits passes through the sun's center. Thus we have one point of every axis, line of nodes, and plane of every orbit of the primary planets.

To obtain the longer axis we have only to measure the planet's distance from the sun when in aphelion and in perihelion. These distances added together make the longer axis of the orbit. The perihelion distance being known, we readily obtain the eccentricity, hence the shorter axis, and from these the entire ellipse in magnitude. The point at which a planet passes from north to south of the ecliptic is one point in its line of nodes, the sun's center is another, and these determine the direction of the line of nodes. The inclination of plane of the planet's orbit to the ecliptic is measured by the angle formed between a line drawn from the sun's center perpendicular to the line of nodes in the plane of the ecliptic, and one perpendicular to the same line at the same point, but lying in the plane of the planet's orbit. The elevation therefore of the planet above the ecliptic, when 90° degrees from the node, will be the *angle* of *inclination.* Having the line of nodes and inclination, we can draw the plane. Having the perihelion point, longer axis and eccentricity, we can construct and locate the elliptic orbit, and having the moment of perihelion passage, we can trace the planet in its future movements.

The elliptical theory being adopted and extended to all the known planets successfully, it became manifest to the searching genius of Kepler that there existed too many common points of resemblance between these revolving orbs not to involve some common bond which united them into a scheme of mutual dependence. They all revolved in elliptical orbits. These orbits had one common focus, the sun. The lines of nodes and principal axes intersected in the sun. They all obeyed the same law in their revolution in their orbits, and Kepler now undertook the task, almost hopeless in its character, of discovering some bond of union which might reduce a multitude of now isolated worlds to an orderly and dependent system.

This problem occupied the mind of Kepler for no less than nineteen years. He examined carefully all the elements of the planetary orbits, and finally selected the *mean distances* and *periodic times* as the objects of his special investigation. He found that the *periods* of revolution increased as the planet was more remote from the sun, but certainly not in the exact ratio of the distance. Thus—

The mean distance of the earth is . .	95,000,000 of miles.
Its period of revolution, ...	$365\frac{1}{4}$ days.
Mean distance of Mars, ...	142,000,000 of miles.
Period of revolution, ...	687 days.

In case the distances and periodic times were exactly proportional, we should have $\frac{142}{95}=\frac{687}{365}$. But $\frac{142}{95}=1.5$ nearly, while $\frac{687}{365}=1.9$ nearly. Finding that no simple proportion existed between these quantities, Kepler broke away from the ratios of geometry, which up to his own era had almost exclusively been employed in all astronomical investigations, and conceived the idea that the hidden secret might be found in proportions existing between some powers of the quantities under consideration. He first tried the *squares*, or simple products of the quantities by themselves. Here he was again unsuccessful. He now rose yet higher, and examined the relations of the *cubes* of the periods and distances. But no proportion was found to exist among these *third powers*. At length he was led by some influence, he knew not what, as he says, to try the relation between the squares of the periods and the cubes of the distances, thus, $\frac{687}{365}\times\frac{687}{365}$, and $\frac{142}{95}\times\frac{142}{95}\times\frac{142}{95}$; and here lay the grand secret, for if any one will perform the operations above indicated, and square the periods of revolution, and cube the mean distances, he will find the above quantities to be equal to each other, or, in other language, he will find *the squares of the periodic times exactly proportional to the cubes of the mean distances.*

This is called the *third law* of Kepler, and is perhaps the grandest and most important of all his wonderful discoveries. Through its power the worlds are all linked together. The satellites of the planets revolve in obedience to its sway, and even those extraordinary objects, the revolving double stars, are subjected to the same controlling law. It resolves at once the most difficult problems involved in the solar system, affording a simple method of determining the mean distances of *all* the planets, by measuring the mean distance of any one planet, and by observing the periods of revolution.

As we have already seen, the periodic times are readily determined from noting the days and fractions of days which elapse from the planet's passage through its node until it returns to the same node again. This, in case the line of nodes remained absolutely fixed, would give the time of revolution precisely, and a slight correction suffices to correct the error due to the movement of the nodes. The determination of the mean distance of the earth, then, becomes the key to a knowledge of all the planetary distances, from which flows the absolute magnitudes of the planets and their densities.

It is not, then, surprising that Kepler, seeing the grandeur of the consequences flowing from the great discovery, should have given utterance to his feelings in language of the most lofty enthusiasm.

With the knowledge of the three laws discovered by Kepler modern astronomy commenced a career of wonderful success. We shall find, hereafter, that even these great laws of Kepler are but corollaries to a higher law yet remaining to be developed, but we prefer to follow out the order of examination and development already commenced.

We resume our discussion of the planet under examination. The changes in the apparent diameter of Mars must, of course, be very great. When in opposition to

MARS. AUG. 30. 8 H. 55 M. 1845.

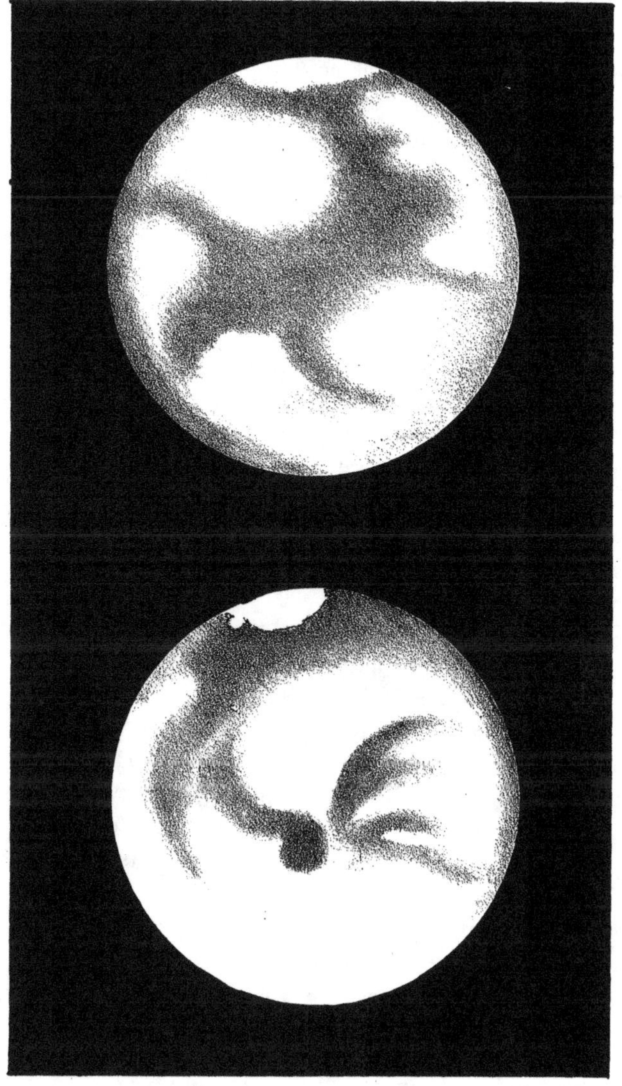

MARS, CINCINNATI OBSERVATORY
AUG 5TH 1845.

the sun, or on a line joining the sun and earth, Mars is only forty-seven millions of miles from our planet, while, on reaching his conjunction with the sun, this distance is increased by the entire diameter of the earth's orbit, or 196 millions of miles. When in opposition Mars shines with great splendor, presenting to the eye, as shown by the telescope, a large and well-defined disk, with a surface distinctly marked with permanent outlines of what have been conjectured to be continents and oceans. The polar regions are distinguished by zones of brilliant white light, which, in consequence of their disappearance under the heat of summer, and their reappearance as the winter comes on, have been considered as due to snow and ice. I have examined these snow zones with the great refractor of the Cincinnati Observatory, under peculiarly favorable circumstancs. To illustrate the mode of observation employed in the determination of the period of rotation of Mars on its axis, and the power of the telescope in the revelation of the physical constitution of this planet, I append some account of Maedler's observations, made in 1830, and also of those made at the Cincinnati Observatory in 1845:—

The last opposition of Mars, which occurred on the 20th August, 1845, furnished a fine opportunity for the inspection of the irregularities of its surface. When in opposition the planet rises as the sun sets, and the earth and planet are in a straight line, which, by being prolonged, passes through the sun. As the orbit of Mars incloses that of the earth, it will be seen from a little reflection that when Mars is in opposition it is nearer to the earth than at any other time, nearer than when in conjunction by the entire diameter of the earth's orbit, or 190 millions of miles. In case the orbits of Mars and the earth were exact circles, the distance between the two planets at every opposition would be the same, but the elliptic figure of the orbits occasions a considerable variation in this distance, and the least distance possible between the earth and Mars will be when an opposition occurs at the time that the earth is furthest from the sun and Mars nearest to the sun. Such was approximately the relative positions of the planets in 1845, and their distance was then less than it can be again for nearly 15 years. During the opposition which occurred in 1830, the earth and Mars held nearly the same relative positions. The planet was observed by Dr. Maedler, the present distinguished Director of the Imperial Observatory at Dorpat, Russia, assisted by Mr. Beer. I have translated the following notices from Schumacher's journal:—

"The opposition of Mars which occurred in the month of September of this year (1830), and at which time this planet approached nearer the earth than it will again for 15 years, induced us to observe the planet as often as the clouds would permit, in order to determine the position and figure of its spots; their possible physical changes, and especially the time of revolution on its axis. The telescope employed was a Franenhofer Refractor, 4½ feet focus.

"The opposition occurred on the 19th September, and the nearest approach to the earth (0,384) on the 14th of the same month. In all succeeding oppositions up to 1845, this distance amounts to 0.5, and even up to 0.65 (the unit being the mean distance of the earth from the sun). On account of the accurate definition of the instrument, we were able to employ a power of 300 generally, and never less than 185. With low magnifying powers, the greatest diameter was determined to be a little less than 22″. Our observations extended from the 20th September to the 20th October, during which time 17 nights, more or less favorable, occurred, and all sides of Mars came into view. Thirty-five drawings were executed. It was not thought advisable to apply a micrometer, as the thickness of the lines would have produced greater errors in such minute measures than those arising from a careful estimation by the eye. The drawings were invariably made with the aid of the telescope. Commonly a little delay was had, till the undetermined figure of the spots visible at the first glance separated themselves (to the eye) into distinct portions.

* * * * * * *

"On the 10th September a spot was seen so sharp and well defined, and so near the center of the planet, that it was selected to determine the period of rotation. On the 14th September it retrograded from the eastern hemisphere, through the center to the western hemisphere, in the course of three hours. Its figure unaltered during four days, and its regularity as to rotation left no doubt of its identity and permanence.

"In the course of 2¼ hours Mars exhibited an entirely different appearance. The spot (already alluded to) was near the western disk of the planet. On the 16th it was again observed, and the period of revolution deduced. It was invisible up to the middle of October, appearing only in the day time on the side of the planet next to the earth. It was first observed again on the 19th October, and the disk of Mars showed itself with uncommon sharpness. On the southern border of the principal spot two red spots were seen, resembling a ruddy sky on the earth. They appeared fainter an hour after, and although they again seemed brighter they were never again seen red. We also observed a faint spot near the principal one, which was never after visible.

* * * * * * *

"The observations from the 26th September to the 5th October showed to us some very dark spots, which in zone-formed extensions showed a strong contrast to the brightly illuminated surfaces free from spots. A fragment of one of these spots was at the north end distinct and broad, while at the south end it was so small as to be seen with difficulty. Between the pole and the principal spot, there was seen a broad stripe, of less shade, while the northern hemisphere was almost entirely free from spots. Bad weather interrupted the observations from the 5th to the 12th October.

"On the 13th, a spot appeared for the first time again, but so near the western disk that we recognized its return only on the 14th.

"More accurate observations were had on the 19th and 20th October, when this spot passed the middle of Mars, which movement was observed with all accuracy, and hence a new determination of the period of revolution. Computation gave the magnitude of the invisible part of Mars on the 13th October=0.06, on the 20th, 0.08, of the radius of Mars.

"From the beginning of the observations there was seen at the south pole, always with great distinctness, a white, glittering, well defined spot, which has long been observed, and is called the '*snow zone.*' During the observations it continually diminished up to the 5th of October. Here an increase commenced, yet very slow. On the 10th September we estimated it, =.110; 5th Oct.=.110, and 20th Oct =.115 of the diameter of Mars.

"In case we adopt Herschel's determination of inclination and position of the axis of Mars, with reference to its orbit, the south pole of Mars on the 14th of April, 1830, must have had its equinox, and on the 8th September, its summer solstice. The smallest diameter of the '*snow zone*' occurred on the 27th day after the summer solstice, a time which corresponds to the last half of July on the northern hemisphere of the earth, at which time it is well known we have the greatest heat.

"Preceding observers in oppositions, where the pole was further from the maximum temperature, have seen the 'snow zone' much larger, although nearly all regard it as changeable in size. These facts seem to sustain the hypothesis of a covering of snow."

As a further confirmation of this hypothesis, we subjoin the following computations, by the same persons. The previous determinations of the elder Herschel are taken as the basis of the calculations. This white polar region is now distinctly visible, and seems to be accounted for in no other way. Comparing the various seasons in Mars, Maedler finds as follows: —

"Duration of Spring,	N. Hemisphere,	. .	191½	Mars' days.
"	Summer,	" . . .	180	"
"	Autumn,	" . . .	149½	"
"	Winter,	" . . .	147	"

"Adding spring and summer together, and fall and winter, we have —

"Duration of Summer in N. H. to S. H., . . . as 19 to 15
"Intensity of sun's light in N. H. to S. H., . . as 20 to 29

"Uniting these two proportions, and assuming that heat and light are received in equal ratios, it will follow that the south pole, by the greater intensity of solar heat, is more than compensated for the shortness of its summer. But since for the winter the proportion of 20 to 29 is reversed, so will the winter of the south pole, not only on account of longer duration of cold, but also from its greater intensity, be far more severe than in the north pole.

"Herewith agree the facts that preceding observers have not lost sight of the 'snow zone' of the south pole, even when the pole became invisible, whence it follows that it must extend from the pole 45° degrees and even further, while we, under like circumstances, could not discover any such appearance on the north side of Mars. On the contrary, the brightness of this portion was exactly like that of the other parts of the disk."

The conclusions reached by the German astronomers, as above, were confirmed in the fullest manner by the observations made at the Cincinnati Observatory during the opposition of 1845. I will here record some singular phenomena connected with the "snow zone," which, so far as I know, have not been noticed elsewhere.

On the night of July 12th, 1845, this bright polar spot presented an appearance never exhibited at any preceding or succeeding observation. In the very center of the white surface was a *dark spot*, which retained its position during several hours, and was distinctly seen by two friends, who passed the night with me in the observatory. It was much darker and better defined than any spot previously or subsequently observed here, and, indeed, after an examination of more than eighty drawings of the surface of this planet by other observers at previous oppositions, I find no notice of a dark spot ever having been seen in the bright snow zone. On the following evening no trace of a *dark spot* was to be seen, and it has never after been visible.

Again, on the evening of August 29th, 1845, the snow zone, which for several weeks had presented a regular outline, nearly circular in appearance, was found to be somewhat flattened at the under part, and extended east and west so as to show a figure like a rectangle, with its corners rounded. On the evening of the 30th August I observed, for the first time, a small bright spot, nearly or quite round, projecting out of the lower side of the polar spot. In the early part of the evening the small bright spot seemed to be partly buried in the large one, and was in this position at 8h. 55m., when the drawing, No. 1, was made. After the lapse of an hour or more, my attention was again directed to the planet, when I was astonished to find a manifest change in the position of this small bright spot. It had apparently separated from the large spot, and the edges of the two were now in contact, whereas when first seen they overlapped by an amount quite equal to one-third the diameter of the small spot. On the following evening I found a recurrence of the same phenomena. In the course of a few days the small spot gradually faded from the sight and was not seen at any subsequent observation. Should Herschel's hypothesis be admitted, that the bright zone is produced by snow and ice near the pole of the planet analogous to what is known to exist at the poles of the earth, these last changes may be accounted for, by supposing the small bright spot to have been gradually dissipated by the heat of the sun's rays.

Its apparent projection over the boundary of the large snow zone may have been merely optical, and the separation may have been occasioned by seeing the two objects in such position as to prevent the one from being projected on the other. Such change may have been produced by the rotation of Mars on its axis in the space of a few hours.

To determine the exact period of rotation of Mars, Sir William Herschel instituted a series of observations in 1777, which were followed by others during the opposition of 1779. From the first series an approximate period of rotation was obtained, and by uniting the observations of 1777 and those of 1779, and using 24h. 39m. as the approximate period of rotation, Herschel made a further correction, and fixed the rotation at 24h. 39m. 21.6s.

Maedler's determination, in 1830, gave, for a final result, 24h. 37m. 10s., which, in 1832, was corrected and fixed at 24h. 37m. 23.7s.

In 1839 Maedler reviewed Herschel's observations, from whence his first results were deduced, and discovered that after introducing the necessary reduction, the discrepancy of two minutes might be reduced to two seconds, by giving to Mars one more rotation on its axis, between the observations of 1777 and 1779, than Herschel had employed.

In 1845, when Mars again occupied the same relative position that it had done in 1830, it was too far south for observation at Dorpat.

By combining Maedler's observation, made at Berlin, 1830, September 14, 12h. 30m., with one made at the Cincinnati Observatory, 1845, August 30, 8h. 55m., making the corrections due to geocentric longitude, phase, and aberration, I find the period of rotation to be 24h. 37m. 20.6s., differing by only two seconds from Maedler's period as last corrected.

It is generally believed that Mars is surrounded by an atmosphere which in many respects resembles our own.

In case this be true, we may anticipate the existence of belts of clouds, and occasional cloudy regions, which would modify the outline of the great tracts of sea and land, and would account for the rapid changes which are sometimes noticed in the surface of the planet.

The axis of the planet is inclined to its orbit (as may readily be deduced from the rotation of the spots) under an angle of a little more than 30°, hence the variations of climate and the changes of season in Mars will not be very unlike those which mark the condition of our own planet. Indeed, there are many strong points of resemblance in the planetary features of the earth and this neighboring world. The planes of their orbits are but little inclined to each other, a little less than 2°. Their years are not widely different when we take into account the vast periods which distinguish some of the more distant planets.

The seasons ought to be nearly alike, and the length of day and night, as determined by the periods of rotation of the two worlds, is nearly the same. In case the great geographical outlines are alike, and seas and continents really diversify the surface of Mars with an atmosphere and clouds, the two worlds bear a strong resemblance to each other.

The actual diameter of Mars is only 4,100 miles, or a little more than half the diameter of our earth, while its volume is not much greater than one-tenth part of the volume of our planet.

To the inhabitants of Mars (if such there be) the earth and moon will present a very beautiful pair of indissolubly united planets, showing all the phases which are presented by Mercury and Venus to our eyes, the two worlds never parting company, and always remaining at a distance of about one quarter of one degree, or about half the moon's apparent diameter.

The amount of heat and light received from the sun by Mars is about one half of that which falls on the earth; and in case the planet were placed under the identical circumstances which obtain on earth, the equatorial oceans even would be solid ice. This, we have every reason to believe, is not the case, and hence we are induced to conclude, as in other cases, that the light and heat of the sun are subjected to special modifications, by atmospheric and other causes, at the surfaces of each of the worlds dependent on this great central orb.

The reddish tint which marks the light of Mars has been attributed by Sir John Herschel to the prevailing color of its soil, while he considers the greenish hue of certain tracts to distinguish them as covered with water. This is all pure conjecture, based upon analogy and derived from our knowledge of what exists in our own planet. If we did not know of the existence of seas on the earth, we could never conjecture or surmise their existence in any neighboring world. Under what modification of circumstances sentient beings may be placed, who inhabit the neighboring worlds it is vain for us to imagine.

It would be most incredible to assert, as some have done, that our planet, so small and insignificant in its proportions when compared with other planets with which it is allied, is the only world in the whole universe filled with sentient, rational, and intelligent beings capable of comprehending the grand mysteries of the physical universe.

CHAPTER VI.

THE ASTEROIDS: A GROUP OF SMALL PLANETS, THE FIFTH IN THE ORDER OF DISTANCE FROM THE SUN.

THE INTERPLANETARY SPACES.—KEPLER'S SPECULATIONS.—GREAT INTERVAL BETWEEN MARS AND JUPITER.—BODE'S EMPIRICAL LAW.—CONVICTION THAT A PLANET EXISTED BETWEEN MARS AND JUPITER.—CONGRESS OF ASTRONOMERS.—AN ASSOCIATION ORGANIZED TO SEARCH FOR THE PLANET.—DISCOVERY OF CERES.—LOST IN THE SOLAR BEAMS.—REDISCOVERED BY GAUSS.—THE NEW ORDER DISTURBED BY THE DISCOVERY OF PALLAS.—OLLER'S HYPOTHESIS.—DISCOVERY OF JUNO AND VESTA.—THE SEARCH CEASES.—RENEWED IN 1845.—MANY ASTEROIDS DISCOVERED.—THEIR MAGNITUDE, SIZE, AND PROBABLE NUMBER.

THE worlds thus far examined in our progress outward from the sun have been known from the earliest ages. Those constituting the group under consideration, called *asteroids*, have all been discovered since the commencement of the present century.

The circumstances attending the discovery of

CERES, ① OF THE ASTEROIDS are replete with interest, and demonstrate the power of the conviction in the human mind that, in the organization of the physical universe, some systematic plan will be found to prevail. In drawing to a scale the solar scheme of planetary orbits, it was readily observed that the distances of the planets from the sun increased in a sort of regular order up to the orbit of Mars. Here, between Mars and Jupiter, there was found a mighty interval, after which the order was restored as to the planets beyond the orbit of Jupiter.

As early as the beginning of the seventeenth century,

Kepler, whose singular genius was captivated by mystical numbers and curious analogies, conjectured the existence of an undiscovered planet in this great space which intervened between Mars and Jupiter. The thought thus thrown out required no less than two hundred years to take root and yield its legitimate fruit. The discovery of a planet beyond the orbit of Saturn, by Sir William Herschel, in 1781, greatly strengthened the opinions based on the orderly arrangement of the interplanetary spaces; and the German astronomer, Bode, by the discovery of a curious relation, which seemed to control the distances of the planets, gave additional force and power to the conjecture of Kepler. This law is a very remarkable one, and although no explanation could be given of it, was verified in so many instances as almost to force one to the conclusion that it must be a law of nature. We present the law in a simple form. Write the series—

	0,	3,	6,	12,	24,	48,	96, &c.
add	4	4	4	4	4	4	4, &c.
sum	4	7	10	16	28	52	100, &c.

Now, if *ten* be taken to represent the distance of the earth from the sun, the other terms of the series will represent with considerable truth the distances of the other planets, as we will readily perceive, thus:—

Mercury.	Venus.	Earth.	Mars.		Jupiter.	Saturn.	Uranus.
4	7	10	16	28	52	100	196

The true distances are roughly as under:—

3.8	7.2	10	15.2		52	95.3	191.8

It is thus seen that the actual distances of the planets agree in a most remarkable manner with those obtained by the application of *Bode's Law*, and as no planet was yet known to fill the distance (28) between Mars and Jupiter, it required very little devotion to the analogies of nature to create in any mind a firm belief in the existence of an unknown planet.

The German astronomers, at the close of the last century, took up the matter with earnest enthusiasm, and in the year 1800 a congress or convention of astronomers was assembled at Lilienthal, of which M. Shroeter was elected president, and Baron De Zach perpetual secretary. It was agreed to commence a systematic search for the unknown planet, by dividing the belt of the heavens near the sun's path, called the zodiac (and within whose limits all the planetary orbits are confined), among twenty-four astronomers, who with their telescopes should search for the object in question.

It was manifest that the unknown planet must be very small, too small to be visible to the naked eye, otherwise its discovery must have been long since accomplished. It might, however, prove to be large enough to exhibit a planetary disk in the telescope, in which event a simple search was all that was required. If, however, it should be too diminutive to show a well defined disk in the telescope, then another method of examination would be required. The planet could only be detected by its motion among the fixed stars. This, indeed, is the way in which all the old planets had been discovered; but while the naked eye takes in at the same time a large portion of the celestial sphere, the telescope is extremely limited in its *field of view*, rendering the search laborious and difficult. Were it possible, however, to make an exact chart of all the stars in a given region of the heavens, to-night, if an examination on to-morrow night of the same region should show a *strange star* among those already charted, this stranger might with some probability be assumed to be a planet.

A few hours of patient watching would show whether it was in motion, and a few nights of observation would reveal its rate of motion.

Such was the mode of research adopted by the society of planet-hunters. The system thus adopted had not been pursued but a few months when a most signal success crowned the effort. On the night of the 1st January, 1801, Piazzi, of Palermo, in Sicily, observed a star in the constellation Taurus, which he suspected to be a stranger. On the following night (having fixed its position anew with reference to the surrounding stars), he found it had changed its place by an amount so large that its real motion could not be doubted. The star was found to be retrograding, or moving backward, and this continued up to the 12th January, when it became stationary. It was soon after lost in the rays of the sun, thus becoming invisible, before any considerable portion of its orbit had been observed, and before Piazzi could communicate his discovery to any member of the society.

Piazzi not considering it possible that a planet which had remained hidden from mortal vision from its creation could be discovered with so little effort as had thus far been put forth, conceived that the moving body which he had discovered was a comet, but the intelligence having been communicated to the society, Bode promptly pronounced this to be the long sought planet, an opinion in which he was sustained by Olbers and Buckhardt, Baron de Zach, and Gauss, and I know not by how many other members of the society.

It now became a matter of the deepest interest to re-discover this stranger after its emergence from the sun's rays, a task of no little difficulty, as we will see by the slightest reflection. The star had been followed through only about 4° of its orbit, and on this slender basis it seemed almost impossible to erect a superstructure such as might conduct the astronomer to the point occupied at any given time by this almost invisible world. We shall see hereafter that this most astonishing feat was successfully accomplished by the German mathematician and astronomer, Gauss, then quite a young man, and who, in this early effort, gave evidence of that high ability for which he became afterward so greatly distinguished.

Ceres being re-discovered, and closely observed, the data were soon obtained for the exact computation of the elements of its orbit, when it was found to occupy, in the planetary system, the precise position which had been assigned to it fifteen years before by Baron de Zach, in accordance with the indications of the curious empirical rule, already presented, known as *Bode's law*.

The harmony of the system was thus fully established, the missing term in the series was now filled. The vast interplanetary space between Mars and Jupiter was the real locality of a discovered world, whose existence had been conjectured by Kepler two hundred years before, and whose discovery, by combined systematic and scientific examination, constituted the crowning glory of the age. True, the new planet was exceedingly small when compared with any of the old planets, yet it acknowledged obedience to the great laws established by Kepler, revolving in an elliptical orbit of very considerable eccentricity, and sweeping round the sun in a period of about *four years* and *nine months*, and at a *mean distance* of about 263 *millions of miles.*

The telescope yielded but little information as to the absolute magnitude and condition of Ceres. Its diameter has been measured by various astronomers but the results are so discordant that but little confidence is to placed in them. It cannot, probably, exceed 1,000 miles, and may be much less. It is supposed to be surrounded by an extensive atmosphere, but the evidence of this is not very reliable. Under favorable circumstances, and with a powerful telescope, a disk can sometimes be seen, but for the most part Ceres presents the appearance of a star of about the eighth magnitude.

Such was the condition of astronomy, affording to those interested cause for high gratification in the now known orderly distribution of the planetary orbs, when an announcement was made which was received with profound astonishment, as it at once introduced confusion precisely at the point in which order had been so lately restored. This was the discovery of another small planet, by Olbers of Bremen, revolving in an orbit nearly equal to that occupied by Ceres. Computation and observation united in fixing, beyond doubt, this most extraordinary discovery, and the new and anomalous body received the name of Pallas. The exact elements of the orbit of Pallas having been determined, it was found that a very near approximation to equality existed between the mean distances and periods of Ceres and Pallas, as we find below:—

Ceres' period of revolution, ...	1,682.125 days.
Pallas' " "	1,686.510 "
Ceres' mean distance, ...	262,960,000 miles.
Pallas' " "	263,435,000 "

Here we find the mean distances and periods so nearly equal, that in case the planes of the orbits of the two planets had chanced to coincide, these two worlds might travel side by side for a long while, and at a distance from each other only about double the distance separating the earth from her satellite. The distance between Mars and Ceres is no less than 120 millions of miles. The distance from Ceres orbit to that of Jupiter is more than 280 millions of miles, and yet here are two planets which may approach each other to within a distance less than half a million of miles.

It is true, the eccentricities of the orbits differ greatly, and the inclinations of their orbital planes is also very great, so that Pallas, by this inclination, is carried far beyond the limits within which the planetary excursions north and south of the ecliptic had been previously confined, yet a time would come in the countless revolutions of these remarkable worlds when each would fill, at the same time, points of the common line of intersection of their orbital planes, and these two points, owing to the revolutions of the perihelion, might, possibly, at some future period, come to coincide.

In case these speculations were within the limits of the probable, and if it were permitted to anticipate in the future, the *possible collision* or union of these minute planets, a like train of reasoning, running back into the past, would lead to the conclusion that in case their revolution had been in progress for unnumbered ages, there was a time in the past when these two independent worlds might have occupied the same point in space, and hence the thought that possibly they were fragments of some great planet, which, by the power of some tremendous internal convulsion, had been burst into many separate fragments. This strange hypothesis was first propounded by Dr. Olbers, and has met with more or less favor from succeeding astronomers, even up to the present day, as we shall see hereafter.

True or false, it soon produced very positive results, for it occasioned a renewal of the research which had been discontinued after the discovery of Ceres, and in a few years two more planets were added to the list of asteroids. The search was long continued, and it was not until the end of fifteen years that Olbers and his associates became satisfied that no more discoveries could be expected to reward their diligence. Thus it became a received doctrine that in case a large planet had been rent asunder by some internal explosive power, it had been burst into four pieces, and that no other fragments existed sufficiently large to be detected even by telescopic power.

This opinion prevailed up to December, 1845, when the astronomical world was somewhat startled by the announcement of a new asteroid, discovered by Henke, of Dreisen. This event awakened attention to this subject, and a new generation of observers entered the field of research, whose efforts have resulted in revealing a large group of small planets, of which no less than fifty-five have already been discovered, and their orbits computed.

The theory of the disruption of one great planet as the origin of the asteroids has been revived and extensively discussed, but thus far no satisfactory conclusion has been reached. So strangely are the orbits of these bodies related to each other that, in case they all laid on the same plane, they would in some instances intersect each other, exhibiting relations nowhere else found in the solar system. None of the asteroids are visible to the naked eye, nor are they distinguishable from the stars with the telescope, except under the most favorable circumstances.

When carefully watched some of them exhibit rapid changes in the intensity of their light, sometimes suddenly increasing in brightness, and again as rapidly fading out. These changes have been accounted for on the supposition that these worlds are indeed angular fragments, and that, rotating on an axis, they sometimes present large reflective surfaces, and again angular points, from whence but a small amount of light reaches the earth.

As the stars of the smaller magnitudes are becoming more extensively and accurately charted, their places being determined with great precision, we may anticipate a large increase in the number of known asteroids during the remainder of the current century, and so forward; for if so great a multitude has already been revealed almost without effort, and nearly by accident, what must be the result when a systematic scheme of examination shall have been executed, based on an accurate knowledge of the places of all the stars down to the twelfth magnitude? We have just ground for supposing that there are thousands of these little worlds revolving in space.

CHAPTER VII.

JUPITER, ATTENDED BY FOUR MOONS, THE SIXTH PLANET IN THE ORDER OF DISTANCE FROM THE SUN.

ARC OF RETROGRADATION.—STATIONARY POINT.—DISTANCE OF THE PLANET DETERMINED.—PERIODIC TIME.—SYNODICAL REVOLUTION GIVES THE SIDEREAL. —SURFACE OF JUPITER AS GIVEN BY THE TELESCOPE.—PERIOD OF ROTATION. DIAMETER.—VOLUME.—MEAN DISTANCE.—AMOUNT OF LIGHT AND HEAT.—FIGURE OF JUPITER.—EQUATORIAL AND POLAR DIAMETERS.—DISCOVERY OF THE FOUR MOONS BY GALILEO.—EFFECT ON THE COPERNICAN THEORY.—JUPITER'S NOCTURNAL HEAVENS.

THE SATELLITES OF JUPITER.—HOW DISCOVERED.—THEIR MAGNITUDE.—FORM OF THEIR ORBITS.—PERIOD OF REVOLUTION.—ECLIPSES.—TRANSITS.—OCCULTATIONS.—VELOCITY OF LIGHT DISCOVERED.—TERRESTRIAL LONGITUDE.—ROTATION OF THESE MOONS ON AN AXIS.

IN passing from the diminutive asteroids to the magnitude and splendor which distinguish the vast orb which holds the next position in the planetary system, we are the more disposed to adopt the theory that the exceeding disparity now existing in the magnitude of these neighboring worlds is due to the fact that the asteroids are but a few of the fragments of some object in which they were all once united. We shall hereafter present a speculation on this subject which seems entitled to consideration.

The planet Jupiter is one of the five revolving worlds discovered in the primitive ages. Its revolution among the fixed stars is slow and majestic, comporting well with its vast dimensions, and the dignity conferred by four tributary worlds.

Like all the old planets, the ancients had determined with considerable precision the period of revolution of Jupiter, and his relative position among the planetary worlds. The points in his orbit where he becomes stationary, the arc over which he retrogrades, and his period of retrogradation, were all pretty well determined from the early observations.

As we recede to greater distances from the sun, the arc of retrogradation diminishes in extent, while the time employed in describing these arcs must by necessity increase. This will become evident if we recall to mind the cause of this apparent retrogradation. When the sun, earth, and planet, are all on the same straight line, the earth and planet being on the same side of the sun, then the planet is exactly in *opposition*. The earth and planet starting from this line, as the earth moves the swifter in its orbit, at the end of, say, twenty-four hours, the line joining the earth and planet will take a direction such that it will meet the first line exterior to the orbit of the planet, as seen below:—

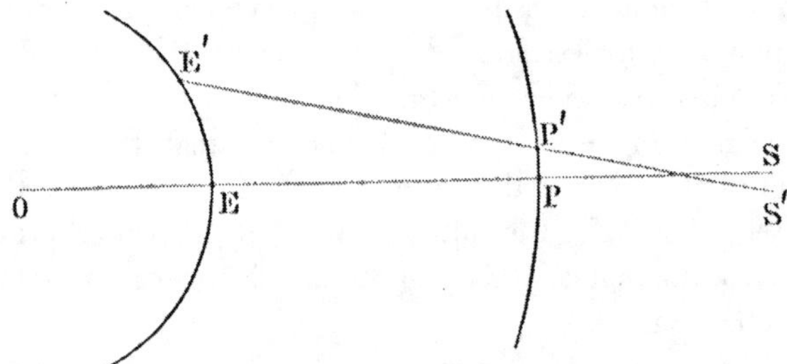

O E P is the line on which the three bodies are found on the day of opposition. At the end of, say, twenty-four hours, the earth arrives at E′ in its orbit, the planet at P′, and then the planet is seen from the earth in the direction E′ P′ S′, whereas on the day previous it was seen in the direction E P S. Thus it appears to have moved backwards from S to S′ among the fixed stars, while in reality it has moved forward in its orbit from P to P′. Admitting the orbits to be circles and the motions to be uniform, it is very easy to locate the places of the earth and planet on successive days after opposition, and joining those places by straight lines, we should soon reach a position in which the lines thus drawn on consecutive days would be parallel. There the planet would appear *stationary* among the fixed stars, and there its *advance* would commence, as is manifest from the figure below:—

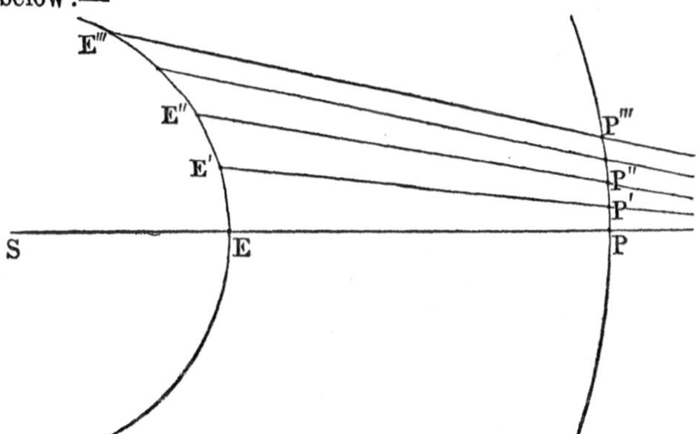

in which S is the sun, E E′ E″ E‴ the successive places of the earth, P P′ P″ P‴ the successive places of the planet. The lines E P and E′ P′ meet on the side opposite the sun, the lines E′ P′ and E″ P″ also meet on the same side, but E‴ P‴ and E″ P″ are parallels, and in P″ the planet becomes stationary, and after passing this point, the earth still advancing, the lines joining the earth and planet meet on the side next the earth, and henceforward the motion of the planet, as seen from the earth, must continue to be direct, until the earth coming round again to occupy the conjunction line, previous to which the stationary point will be passed, and the retrogradation will be commenced.

The distance of any planet from the sun, in terms of the earth's distance, may be obtained from a measurement of the arc of retrogradation in a given time, say twenty-four hours, provided we know the periodic time in which the earth and planet revolve round the sun This will become evident from the figure below—

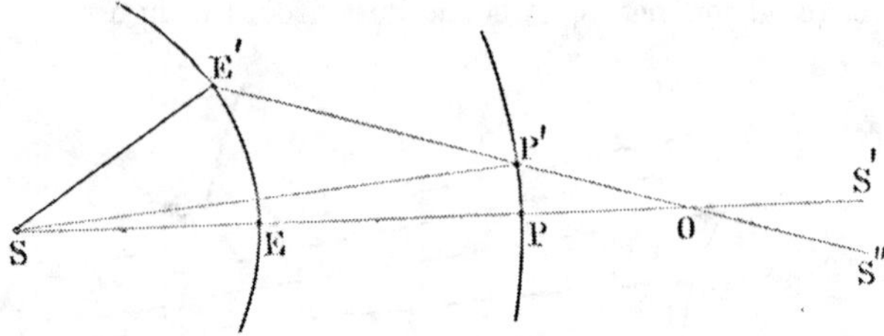

in which S is the sun, E and P the places of the earth and planet on the day of opposition, E′ and P′ their places at the end of twenty-four hours. E E′ and P P′ may be regarded as straight lines, as they are very short in comparison with the entire circumference. As we are supposed to know the periods of revolution of the earth and planet, the distances, E′ E and P P′, are fractional parts of the whole circumference, represented by one, divided by the number of days in the periodic time. The fraction for the earth is $\frac{1}{365.25}$, and for Jupiter it is $\frac{1}{4332.58}$.

In the right-angled triangle, E′ E O, we know the

value of E′ E, and the angle, E′ O E, equal to S′O S″, or the retrogradation of the planet. Hence the other parts become known either by construction or the simplest processes of trigonometry. We thus determine the value of E O, and adding S E, we have the value of S O. Then in the triangle S P′ O, we have the side S O, just determined, also the angles P′ S O and P′ O S. Hence we can construct the triangle, or compute by trigonometry the other parts. Thus S P′, the planet's distance, becomes known.

In case the periodic times were accurately known, and the orbits were exact circles, this mode of determining the distance of a superior planet would be sufficiently exact, but by the third of Kepler's laws, which tells us that the squares of the periodic times are proportional to the cubes of the mean distances, we perceive that the entire problem of the planetary distances resolves itself into fixing, with all possible precision, from observation, the periods of revolution, and then in obtaining the exact distance of any one of them.

We have already stated that the interval elapsing from the passage of a planet from one side of the ecliptic to the other, up to the same again, gives the period of revolution, in case we correct for the various changes which may take place from one node-passage to the next. This, however, in the case of a planet like Jupiter, whose orbital plane nearly coincides with the ecliptic, becomes difficult as a matter of observation, and hence some better method must be employed. This is best accomplished by observing the exact time of *opposition*, or the moment when the planet is 180° distant from the sun.

The interval between two such oppositions is called a *synodical revolution*, and in case the earth did not move, would be the planet's period of revolution around the sun. These *synodical revolutions* would be all precisely equal on the hypothesis of circular orbits and equable motions. But as the planetary orbits are elliptical, and hence the motions variable, the synodical revolutions of any planet, as Jupiter, will vary somewhat from each other in duration. If, however, a large number be counted, say, as many as have occurred in a thousand, or even two thousand years, then a mean period is deduced of great accuracy.

This is possible, as we have the oppositions of the old planets, recorded by the ancients with sufficient precision to be employed in such a discussion.

To derive the sidereal revolution from the *synodical* we have only to consider that the two bodies set out from the same right line. The earth's velocity is known; the time required for the earth to overtake the planet is known (the synodical revolution). The velocity or rate of the planet's motion is required. This is readily found by simple proportion. Take the following example. In a mean solar day the earth travels in its orbit 0°.9856. A mean synodical revolution of Jupiter is observed to be equal to 398.867 solar days. But the earth performs its revolution, and comes again to the starting point in 365.256 days, and then must travel for 398.867—365.256=33.611 days before overtaking Jupiter. But in 33.611 days, at the rate of 0°.9856 per diem, the earth will travel about 33°.928, and this is the whole distance made by Jupiter in 398.867 days. Hence, his rate per diem is, $\frac{33^\circ.928}{398.867}=4' 99=4' 59''.2$, and at this rate to travel 360° will require $\frac{360^\circ}{4' 59''.2}$, =4.332d. 14h.

2m., which is the time occupied by Jupiter in performing his revolution around the sun.

These methods of investigation, which are perfectly simple, were employed by the ancients, and used even by Copernicus, Kepler, and others, and furnished the approximate values of the periods and distances employed in the researches of Kepler, whereby he discovered his celebrated laws.

PHYSICAL CONSTITUTION OF JUPITER.—When examined with powerful telescopes the surface of Jupiter is found to be diversified with shades of greater or less depth, forming parallel bands or belts, especially about the equator of the planet.

Upon these belts well-defined breaks, irregularities, and spots are discerned, by means of which it is discovered that the face of Jupiter, visible at any given time, is completely hidden by rotation on an axis, a new face appearing at the end of a little less than *five hours*. This gives a period of axical rotation of 9h. 55m. 49.7s., as the result of investigations similar to those employed in determining the period of rotation of Mars.

When the apparent diameter of Jupiter is accurately measured, and his distance is taken into consideration, we find his actual diameter to be nearly 90,000 *miles*, and his *volume* to be equal to that of 1,281 globes such as our earth.

The dark belts which encircle the equatorial regions of the planet, and which revolve with the globe, show that the axis of rotation is very nearly *perpendicular* to the plane of the orbit.

Thus we have a planet twelve hundred and eighty-one times larger than our earth, rotating on an axis, but little inclined to the plane of its orbit, in less than ten hours

JUPITER

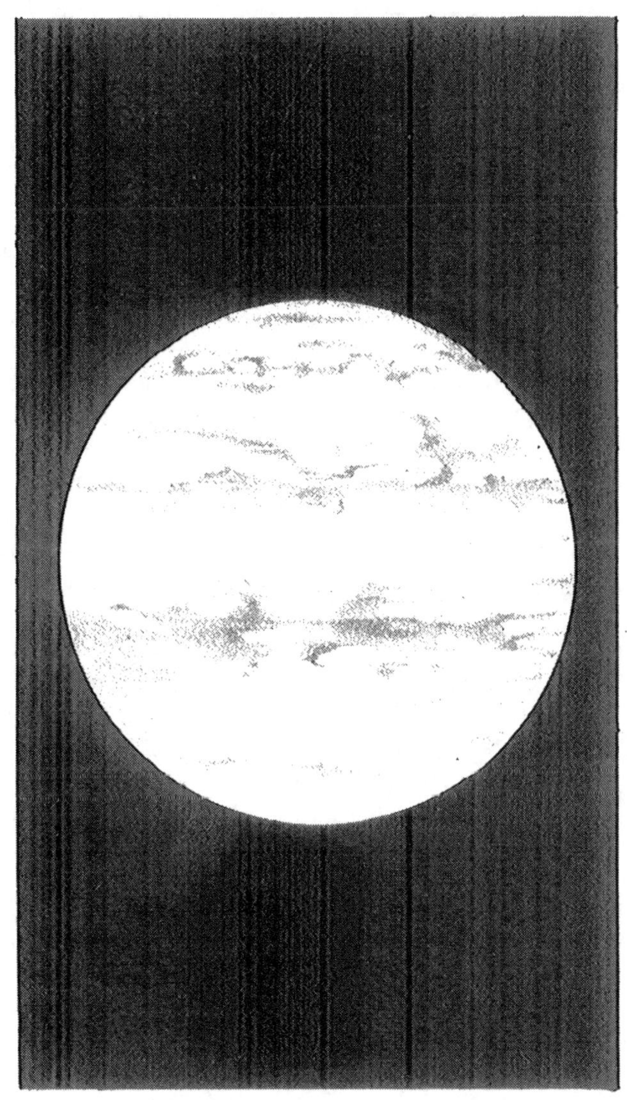

ALLA VISTA, TENERIFFE 1856

of time, and sweeping round the sun in about twelve of our years, at a mean distance of about 485 millions of miles.

The streaks and dark shades which distinguish the equatorial region of Jupiter are by many considered to be belts of clouds floating in the atmosphere of the planet, thus indicating the existence of all the great elements which distinguish the earth. In consequence of the fact that the axis of Jupiter is very nearly perpendicular to the plane of the orbit, the sun will always pour his rays vertically on the equator of the planet. constituting one perpetual summer in all parts of the globe. In case light and heat are governed by the same laws which hold on the earth, the inhabitants of Jupiter will receive from the sun only one twenty-seventh part as much light and heat as falls on the earth. What modifications of heat may be effected by the extensive atmosphere which appears to surround Jupiter it is impossible to conjecture. We may suppose, without reflection, that a world would be only dimly illumined whose sun was reduced to one twenty-seventh part of that which lights our earth. This, however, is not the case, as any one will credit who has ever witnessed the flood of light poured forth from the smallest portion of the sun's disk in emerging from total eclipse. The amount of light which falls on Jupiter far exceeds that which is poured upon the earth on a moderately cloudy day.

When we measure rigorously with the micrometer the figure of Jupiter's disk, we find a marked deviation from the circular outline. This is analogous to the figures of the earth and Mars, and indeed the same peculiarity (of which a satisfactory account will be given hereafter) distinguishes all the planets. In Jupiter the equatorial diameter exceeds the polar by more than *six thousand miles.*

We are indebted to the telescope for the revelation of the highly interesting fact that Jupiter is attended by no less than four moons or satellites, nearly all of them larger than our own moon. These satellites were discovered by Galileo in 1610, soon after he had finished his second telescope, which, as he tells us, cost him incredible pains, and which bore a magnifying power of about thirty times. The discovery of these moons of Jupiter may be regarded as among the most important results of the application of the telescope, if we take into account the then existing condition of astronomical science. The scientific world was just in a transition state. The most honest, intelligent, and powerful minds had already adopted the Copernican theory, but in the universities and other schools of science, as well as in the church, the system of Ptolemy still reckoned among its supporters a host of learned and dignified men. The beautiful miniature of the solar system presented in Jupiter and his moons, as given by Copernicus, could not fail to exert a most powerful influence over all candid and unprejudiced minds. Here was presented to the eye a central orb and about it a scheme of dependent worlds revolving in circular orbits, and with such elegant simplicity as to shame the cumbrous complexity which distinguished the epicyclical theory of the old Greek school. It is not at all surprising that Galileo, the discoverer of this beautiful system, should have become one of the most ardent supporters of the doctrines of Copernicus.

These satellites of Jupiter revolve in orbits whose planes are nearly coincident with the equator of their primary. The exterior, or most distant of the four, revolves in an orbit somewhat inclined to the plane of Jupiter's equator, but the three inner satellites, at every revolution, eclipse the sun to the inhabitants of Jupiter, and are themselves eclipsed in passing through the shadow of their primary. The same phases which mark the revolution of our moon are also exhibited by Jupiter's moons, and the periods of revolution of three of the satellites are so adjusted that one of them must be full when the other two are new.

The nocturnal heavens, as seen from this grand orb, must be inexpressibly magnificent. Besides the same glittering constellations which are seen from earth, the sky of Jupiter may be adorned with no less than four moons, with their diverse phases, some waxing or waning, some just rising or setting, some possibly just entering into or emerging from eclipse.

The whole of this splendid celestial exhibition, sweeping across the heavens, rising, culminating, and setting, in less than five hours of our time. Such are the scenes witnessed by the inhabitants of Jupiter, if such there be.

THE SATTELITES OF JUPITER.—As already stated, these tributary worlds were discovered by Galileo in 1610. On the evening of January the 8th, of that year, having completed his second telescope, capable of bearing a magnifying power of thirty times, he went to his garden to test its quality by an examination of Jupiter. Near the planet he noticed three small stars, nearly in a straight line, passing through the center of Jupiter. He supposed them to be fixed stars, but carefully noted their positions with reference to Jupiter and to each other. On the following night, he remarked that there was a manifest change in the relative places of these stars and the planet which could hardly be accounted for by the motion of Jupiter in his orbit.

Galileo began to suspect the true nature of the stars

which had attracted his attention, and seeing clearly the immense importance of such a discovery, awaited with great impatience the coming of the next evening to confirm his conjectures. Clouds, however, coming up disappointed his hopes, and it was not until the evening of the 14th that he was again permitted to direct his telescope to the planet, when he found, to his great delight, not only the three stars, still in close proximity to the planet, but he also detected a fourth one, whose appearance and position were such that he announced at once the discovery of four moons resembling our own, and revolving about the planet Jupiter as their central orb.

This announcement created the greatest excitement in the astronomical world. Its effect on the old theory of astronomy was at once perceived, and the disciples of Ptolemy determined that they would never believe in the existence of any such pestilent worlds. Some of them actually refused to do so much as look through the tube of Galileo, declaring the whole was a deception, and unworthy the attention of a true philosopher.

The discovery was not the less real because its truth was denied, and to this important addition to the bodies which constitute our system modern science is indebted for some of its most elegant discoveries.

The great distance at which we are compelled to examine these bodies has rendered it difficult to obtain, even with our most delicate instruments, satisfactory measures of the diameters of these satellites. Approximate measures have been obtained from which we learn that the nearest satellite has a diameter of about 2,500 miles, the second, 2,068; the third, 2,377; the fourth, 2,800 miles. We name them in the order of their distances from the primary.

By careful measures of the *elongations*, or greatest distances to which these bodies recede from their primary, the magnitude and form of their orbits have been well determined. The first satellite is thus found to revolve round Jupiter in an orbit nearly circular, whose diameter is 260,000 miles, in a period of 1d. 18h. 28m. The plane on which the orbit lies is inclined to the plane of Jupiter's orbit, under an angle of 3° 05′ 30″, or less, by nearly one-half, than the angle made by the moon's orbit with that of the earth. The smallness of this angle, the nearness of the satellite to its primary, the immense magnitude of the primary and the distance from the sun, combine to produce an eclipse of the first satellite at every revolution, while, in like manner, an eclipse of the sun takes place quite as frequently, from the fact that the shadow of the satellite falls on the planet at every conjunction of the satellite with the sun. These statements are not mere conjectures. They are verified by the telescope, for these eclipses of the satellite and the shadows cast on the primary are distinctly seen from the earth, and furnish the data whereby the periods of revolution are determined with great precision. When Jupiter is in opposition it often occurs that the satellite when on the hither side of the primary, is seen projected on the disk of the planet as a round *bright* spot, while the shadow of the same body may be seen in close proximity as a round *black* spot. Any eye, situated within the limits of this shadow, will witness an eclipse of the sun precisely such as is produced on earth by the shadow of the moon. The passage of the satellite across the disk of Jupiter is called a *transit*. From this position the moon of Jupiter revolves round half its orbit, and then by necessity passes across the cone of shadow cast by the primary in a direction opposite the sun. Here we behold an eclipse of the secondary, as its light is extinguished on entering the shadow, and is only regained after passing beyond the limits of the shadow, thus demonstrating beyond a doubt the fact that, like our moon, these secondaries of Jupiter shine only by reflecting the light of the sun.

In case Jupiter were at rest, it is evident that the observations of these eclipses would give the exact period of revolution of the satellite, which would be precisely the interval from one eclipse to the next. The fact that the earth is in motion would not affect the time of recurrence of the eclipse, for this would be entirely independent of the place of the spectator, provided he sees the disappearance of the satellite at the moment its light is extinguished. It is manifest that the motion of Jupiter in his orbit will change the position of the axis of the shadow, and as the satellite revolves in the same direction in which the shadow advances, it is clear that the time from one eclipse to the next is longer than the true period of revolution of the satellite, by a quantity easily computed from the known orbital velocity of the planet, as may be seen from the figure below, where—

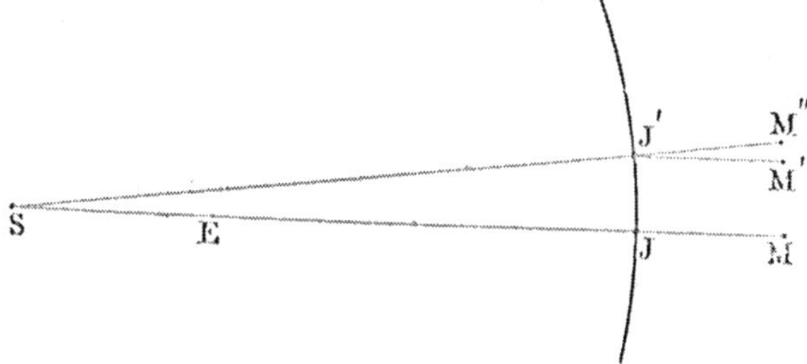

S is the sun's place, E, the earth, J, Jupiter in opposition, M, the satellite in eclipse. At the end of one exact revolution of the satellite, Jupiter has reached J′, the satellite is at M′, but the axis of shadow is now J′ M″, and the center of the eclipse will not occur until the satellite reaches M″, passing over the angle M″ J′ M′. This angle is precisely equal to the angular motion from eclipse to eclipse, a quantity easily determined. The satellite will then revolve 360°, + the angle J S J′, or M′ J′ M″, in the interval from one eclipse to the next, hence the rate per hour becomes known, and gives at once the period of revolution.

Galileo devoted himself for many years to a careful observation of the eclipses of Jupiter's moons, and finally constructed tables whereby these eclipses might be predicted with tolerable precision. His successors devoted much time to the same subject, for a reason we will give hereafter. Long study of these phenomena revealed the curious fact that the interval from one eclipse to the next did not fulfil the prediction based on the foregoing reasoning. The place of the earth seemed in some mysterious way connected with the time at which the eclipse occurred. This may to some appear very reasonable, but, in fact, on the hypothesis that at the moment of the extinction of a luminous object it ceases to be visible, the place of the earth in its orbit or the position of the observer could in no way affect the moment of the satellite's disappearance by entering the shadow of its primary. This will become manifest from a very simple illustration. Suppose the persons in a large circular hall to be gazing on the light of a taper, and the taper is suddenly extinguished by being blown out, every observer will certainly lose the light at the same absolute moment, admitting the fact that the light dies at the instant of extinction to every eye. Let us apply this illustration to the eclipses of Jupiter's moons. They are only seen when the sunlight falls on them. Cut off from them the sunlight by entering the shadow of the primary, and admitting this entrance to be instantaneous, every eye everywhere should lose the light at the same moment of absolute time.

The earth's position in its orbit ought, therefore, to have no effect on the time of the eclipse, and yet it became clearly manifest that the earth's place was in some way connected with certain irregularities in the intervals of these remote eclipses. This matter will be best illustrated from the figure below, in which S represents

the sun, E E′ E″ E‴ the earth's orbit, J Jupiter, and S the satellite. It was found that when the earth was at E, or nearest to Jupiter, the interval from eclipse to eclipse grew longer as the earth receded from Jupiter. At E′ the interval was at a maximum. It now diminished by slow degrees, becoming nearly stationary at E″, then growing shorter, reached a minimum at E‴, after which a slow increase was noticed up to E, and so on in every revolution of the earth in its orbit. Due account, of course, must be taken of the orbital movement of Jupiter. In case the student is ignorant of the explanation of these variations in the synodical revolutions of the moons of Jupiter, he may test his own powers of discovery by a close examination of the facts as above presented. All the satellites gave evidence of the same facts, and the irregularities were found to follow in the same order, reaching their maxima, minima, and stationary points at the same time, or when the earth was at the same point of its orbit.

More than fifty years passed away without any satisfactory explanation of the facts and phenomena above recorded, when, in 1675, Röemer was at length successful in solving the mystery, and found it due to the *progressive motion of light*, which up to this time had been considered by all philosophers as instantaneous in its effects; that is, if a luminous body were created, all eyes, no matter how remotely placed, would see the light at the same moment of time. As the velocity of light, deduced from these investigations, is so enormous, no less than 192,000 miles in one second, we will enter into the explanation somewhat minutely. Suppose a luminous body, as in the figure below, at S, suddenly to be extinguished, the stream of light flowing from the body is

at once cut of, and when the last particles or wave passes a spectator at A, at that moment he will mark the extinction of the light, while to the spectator at B the really extinct luminous body will remain visible until the last particles of the stream of its light pass B, and then the body vanishes to the spectator at B. Suppose the body to thus disappear periodically, A and B will, while they remain stationary, note the intervals from one disappearance to the next to be precisely equal, and the interval, as observed by A, though beginning at an earlier absolute moment of time, will be equal to the interval, as observed at B. Let us now suppose, that after a disappearance, and before the next, A removes to B, it is manifest that the duration or period whose beginning was observed at A, but whose ending was noted at B, will be longer than it was before by an amount of time required for the stream of light to pass from A to B. The reverse would be true if B changed his position to A.

These principles are precisely applicable to the case under consideration. If the earth's orbit were a straight line, with a length equal to A B, the conditions would be identical. The nearly circular figure of the earth's orbit produces the variations already noticed. When the earth is rapidly receding from the source of light, the duration of the synodical revolution of Jupiter's moon will be increased by the time required for the light to pass over the space traversed by the earth during the synodic revolution. This period amounts to some seventeen days for the fourth satellite. But the earth travels some 68,000 miles an hour, or in seventeen days nearly thirty millions of miles, so that the synodical revolution, when longest, will exceed the same period when shortest by an amount equal to the double time required by light to travel 30,000,000 of miles. This difference between the maximum and minimum synodic periods, proved to be about five minutes, and hence it became evident that light must fly at the rate of sixty millions of miles in five minutes or 12,000,000 miles in one minute or 192,000 miles per second.

Should this result appear incredible we shall find

hereafter abundant confirmation of its truth by a train of reasoning and phenomena entirely distinct from what have just been given.

In case light travels with a finite velocity we cannot fail to perceive that this fact will introduce important modifications in all observations designed to fix the places of the heavenly bodies at a given moment of time. Since the earth is sweeping through space with great velocity even this fact will produce a certain displacement in the apparent place of a fixed luminous body. When the body under observation is in motion, the velocity of light being finite, it is clear that the light which falls on the eye of the spectator, and which enables him to see the object, is not the light emitted at the moment the object is seen. Thus, the planet Jupiter is distant from the earth, say, 480 millions of miles. To travel this distance his light must occupy no less than forty minutes, during which time Jupiter has advanced in his orbit about one-third of his own diameter. During the same time the earth has traveled in its orbit a certain distance nearer to or further from Jupiter, which must be taken into account in our effort to fix the absolute position of the planet's center at a given moment. This subject will be resumed when we come to consider the means and instruments employed in astronomical observation.

The satellites of Jupiter have furnished, in their eclipses, the earliest method of resolving the great problem of

Terrestrial longitude.—The position of any place on the earth's surface is determined by fixing its distance from the equator of the earth, north or south, called the latitude, and also its distance east or west of any given meridian line, called the *longitude*. The first of these elements is very readily determined. In case a place is situated on the equator, its latitude is zero, and to any spectator at this place, as we have already shown, the poles of the earth and heavens will lie on the horizon. Leaving the equator and traveling due north along a meridian line, for every degree we go north it is evident the pole of the heavens will rise one degree above the horizon; and when we reach the north pole of the earth, the north pole of the heavens will be on the zenith, or ninety degrees above the horizon.

Thus it appears that the *latitude* of any place is equal to *the elevation of the pole above the horizon of the place*, and to fix the latitude we have only to measure this angle of elevation with a suitable instrument, and apply certain corrections, to be hereafter explained. The problem of the *longitude* does not admit of so easy a solution. To determine accurately longitude at sea is a matter of the highest importance to commerce and navigation, a problem for whose solution maritime nations have in modern times offered large rewards. The safety of a vessel, its crew and cargo, depends on learning by some method its exact position on the surface of the ocean, where there are no permanent objects on our globe to mark its place; and it is only from the celestial sphere that it becomes possible to select fixed objects which may reveal to the mariner the dangers by which he is surrounded.

The latitude, as we have seen, is readily obtained; not so the longitude, which had, up to time of Galileo, been regarded as almost an impossible problem at sea. The great Florentine astronomer saw in the eclipses of the moons of Jupiter the means of solving this highly important problem, and to this end he devoted many years to most diligent and careful observation of these eclipses, with a view to be able to predict their coming, months or even years in advance. We will now explain how these predicted eclipses of Jupiter's satellites, conjoined with their actual observation, may be employed in the determination of *terrestrial longitude.*

As the earth rotates on its axis with uniform velocity, the 360 degrees of the earth's equator are fairly represented by twenty-four hours of time. Thus an hour of time is equal to 15° of longitude, a minute of time is equal to 4′ of longitude, a second of time is equal to 4″ of longitude. The difference of longitude, then, of any two places on the earth's surface is nothing more than the difference of local time, for a mean time solar clock marks 0h. 00m. 00s. when the center of an imaginary sun, moving with the mean or average velocity of the true sun, reaches the meridian of the place in question. A place *west* of the first one will have the center of the mean sun on its meridian *later* by an amount of time equal to the exact difference of longitude. It is clear, then, that if any phenomenon, such as the sudden extinction of a fixed star, could be noted by two observers in different places, each will record the moment of disappearance in his own local time, and an inter-comparison of these records will give at once the difference of longitude between the two stations.

Suppose it were possible to predict that the bright star Vega, in the constellation of the Lyre, would suddenly disappear on the first day of January, 1870, at 0h. 00m. 00s. mean time at Greenwich, England, this fact being known and published, vessels at sea on long voyages, in all parts of the globe, having the star above their horizon, by watching for this phenomenon, and by noting the moment of disappearance in their local time, would determine their longitude from Greenwich. All observers recording the disappearance before the predicted time would be in east longitude, while those recording the same phenomenon *later* than the predicted time would be in *west* longitude, and as many hours, minutes and seconds west as was indicated by their local time.

Now, at sea, very simple methods, as we shall show hereafter, may be employed to obtain the local time, and thus, were it possible to predict a multitude of such phenomena as above recorded occurring every day or two, for years in advance, seamen on long voyages, providing themselves with these predictions, would have the means of fixing their longitude as often as any one of those predicted phenomena could be observed.

The eclipses of the moons of Jupiter are precisely like the phenomenon of the sudden extinction of a star. As these moons shine only by reflected light, the moment they enter the shadow of their primary they vanish from the sight, or are, to all intents and purposes, extinguished; and as these eclipses are constantly recurring at very short intervals, Galileo saw at once the use to which they might be devoted in the resolution of this great problem of *terrestrial longitude*.

Before they could be thus used it became necessary to master completely their laws, so that the moment of eclipse might be accurately predicted years in advance. Though the Tuscan philosopher did not live long enough to perfect and apply his great discovery, his successors in modern times have fully carried out and applied what was so admirably conceived and so carefully commenced.

An attentive examination of the luminosity of Jupiter's moons reveals the curious fact that it is *variable*, increasing and decreasing at regular intervals, equal to the periods of revolution in their orbits, whence it has been inferred by Sir William Herschel and others that each of these satellites rotates (like our moon) upon an axis in the exact time in which it revolves about the primary.

CHAPTER VIII.

SATURN, THE SEVENTH PLANET IN THE ORDER OF DISTANCE FROM THE SUN, SURROUNDED BY CONCENTRIC RINGS, AND ATTENDED BY EIGHT SATELLITES.

The most Distant of the Old Planets.—Its Light Faint, but Steady.—Synodical Revolution.—The Sidereal Revolution.—Advances in Telescopic Discovery.—Galileo announces Saturn to be Triple.—Huygens Discovers the Ring.—Division of the Ring into Two.—Cassini announces the Outer Ring the Brighter.—Multiple Division.—Shadow of the Planet on the Ring.—Belts and Spots.—Period of Rotation of the Planet and Ring.—Disappearance of the Ring Explained.—The Dusky Ring.

SATELLITES OF SATURN.—By whom Discovered.—Eight in Number.—Their Distances and Periods.—Saturn's Orbit the Boundary of the Planetary System, as known to the Ancients.

We now reach, in our outward journey from the sun, the most distant world known to the ancients, revolving in an orbit of vast magnitude, and in a period nearly thirty times greater than that of our earth. Saturn, on account of his immense distance, shines with a fainter light than either of the old planets, though still a conspicuous object among the fixed stars. Its light is remarkably steady, without the scintillations which distinguish the stars, and the brilliant glare which is shown by Venus and Jupiter. There is a yellowish or golden hue to this planet which is not lost when seen through the most powerful telescopes.

Such is the planet Saturn as known to the old astronomers, and as seen by the unaided vision. Its movement among the fixed stars is distinguished by the same phenomena which we have found to exist among all the planets. Being the most remote of all the old satellites of the sun, its stations are the best defined, its arc of retrogradation the shortest, and the period employed in this retrograde movement is longest. From observations made during opposition, and by trains of reasoning identical with those laid down in our examination of Jupiter, the periodic time and mean distance of Saturn are concluded.

Owing to the very slow motion of this planet in its orbit, the earth will pass between it and the sun, or bring it into opposition, in a little over 378 days; that is, Saturn and the earth starting from the same straight line, passing through the sun, the earth makes its revolution, comes up to the starting point, and then overtakes Saturn in about twelve days and three-quarters. The earth's period must then be to that of Saturn as twelve days and three-quarters is to 378, or as one to thirty, roughly.

This determination is a matter of such simplicity that any one, almost without instruments, may make the observations which give the data for the computation. The opposition is observed when Saturn is 180° from the sun, and we have only to count the days from one opposition to the next to obtain the synodical revolution.

Such were the few facts known to astronomy touching this distant orb prior to the discovery of the telescope. The immense multiplication and extension of human vision effected by the invention and improvement of that instrument is in no case more signally displayed than in the successive revelations which have been made in the physical constitution of Saturn, and the extraordinary appendages and scheme of dependent worlds now known to revolve around him.

In 1610, the year in which Galileo first applied the telescope to an examination of the celestial orbs—the year in which he announced the discovery of Jupiter's moons—an examination of Saturn resulted in the strange and anomalous discovery that his disk was not circular, like all the other planets, but elongated, as though two smaller planets overlaid a larger central one extending somewhat to the right and left of the center. This remarkable figure Galileo announced to his astronomical contemporaries under the form of a puzzle produced by a transposition of the Latin sentence—

"Altissimum planetam tergenimum observavi."

"I have observed the most distant of all the planets to be triple."

This mode of presenting the discovery was adopted by the Florentine astronomer to establish his priority, as many of his great discoveries were claimed by some of his opponents, while the truth of all was most obstinately disputed by others. It was urged, even in the case of Jupiter's moons, that these were mere illusions, the offspring of the heated imagination of the ambitious philosopher, and that other eyes could never verify these pretended discoveries. We can readily imagine what must have been the feelings of Galileo when, not many months after the discovery of the triple character of Saturn, he was compelled to acknowledge that, even as seen through his most powerful telescope, the planet was exactly circular, with an outline as sharp and perfect as that of Jupiter. He exclaims, "Can it be possible some demon has mocked me!" He did not live to explain this remarkable change, but he saw the triple form restored, and discovered these periodical transmutations of figure.

Fifty years later, in 1659, Huygens, with more powerful telescopes, discovered the true figure of Saturn, and found the *triple form* seen by Galileo to be produced by the fact that the round planet was encircled by a broad, flat ring of immense diameter, and so situated that the spectator on the earth can never see it in a direction perpendicular to its plane. Hence, although circular in form, the direction of the visual ray gives it an oval or elliptical figure. Huygens distinctly perceived the dark space intervening between the body of the planet and the ring, right and left, which had escaped the eye of Galileo with a less perfect telescope. Hence, the Florentine astronomer only saw the planet elongated, and pronounced it triple. Huygens explained the mysterious change of figure which had so perplexed Galileo, and found it due to the fact that the ring is extremely thin, so thin, indeed, that when the earth chances to hold a place such that the plane of the ring produced passes through the earth and the ring comes to bé presented to the spectator edgewise, not even the telescope of Huygens could discern the fibre of light presented by the rim, or circumference of the ring, when thus located, and to them the disappearance was complete, leaving the planet round, clear, and well-defined.

In 1665, what had hitherto been regarded as one broad, flat ring, was observed to be divided into *two portions* by a dark line, which, under favorable circumstances, was traced entirely round the ring. This discovery was confirmed by the elder Cassini, in 1675, who also discovered the unequal brilliancy of the two rings, the outer one being the brighter. He also was the first to announce the existence of a dark stripe or belt surrounding the equator of the planet. Other discoveries, such as additional belts, the shadow of the planet on the ring, the shadow of the ring on the planet, were successively made, as the powers of the telescope were improved. During the present century many astronomers assert the multiple division of the rings of Saturn, and the evidence is so conclusive, that the existence of dark lines, concentric with the rings, (and like that which severs the two principal rings, cannot be denied,) though there is every reason to believe that these lines are only to be seen occasionally. With the full power of the refractor of the Cincinnati Observatory, defining in the most beautiful manner all the other delicate characteristics of Saturn and his rings, I have never been able to perceive any trace of any other than the principal division.

The bright and dark belts and certain spots, which mark both the surface of the planet and the ring, have furnished the means of fixing the period of rotation of the planet on its axis at 10h. 29m. 16.8s., while the ring revolves on an axis nearly coincident with that of the planet in 10h. 32m. 15s.

If we reflect on the structure and position of Saturn's rings, the phenomena attending its disappearance and reappearance become readily explicable. The plane of the ring produced indefinitely, intersects the plane of the earth's orbit in a straight line. This is called the *line of nodes* of the ring. This line of nodes, remaining nearly parallel to itself, will manifestly move as the ring moves, carried with the planet in its revolution round the sun. During one-half of Saturn's revolution in its orbit the sun will illumine the northern side of the rings, during the other half it will shine on the southern side.

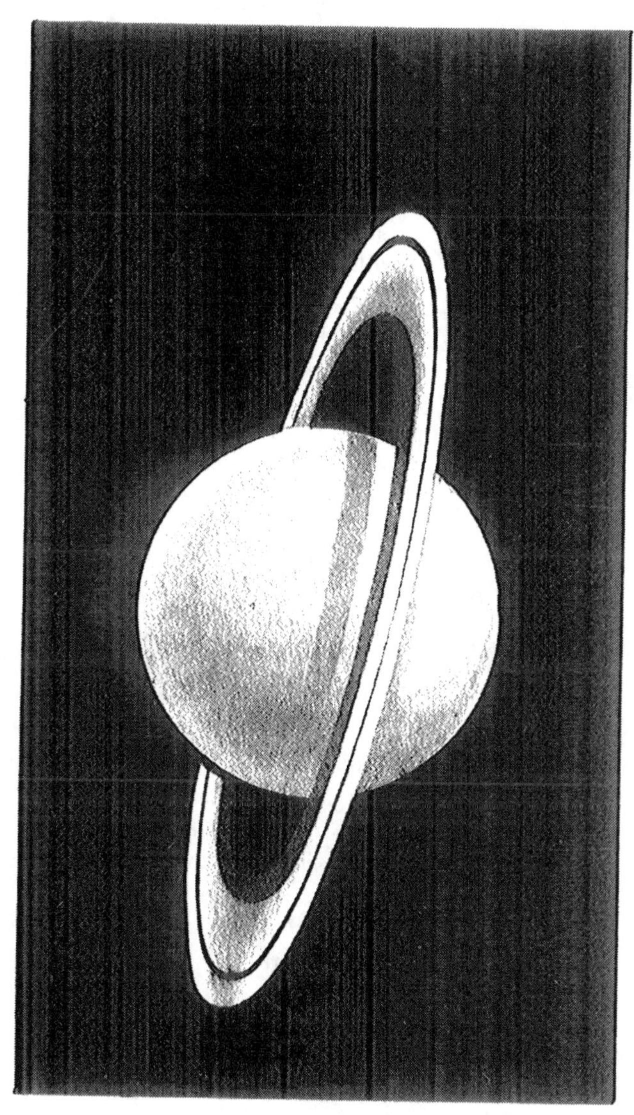

SATURN, CINCINNATI OBS.

Thus the ring, carried by the planet, will finally come into a position such that the sunlight will fall on neither side, but on the edge of the ring only, and when in this position it is manifest that the plane of the ring passes through the sun. If, when in this position, the earth comes between Saturn and the sun, a spectator from the earth's surface will behold the edge of the ring, if visible at all, as a delicate line of light extending beyond the disk of the planet, and passing through its center.

The earth, moving forward in its orbit from opposition of the planet, will pass through the plane of the ring, and upon the non-illuminated side. As Saturn moves very slowly in comparison with the earth, while the plane of the ring is sweeping from the one side of the sun to the other, the earth may pass more than once through the plane of the ring, repeating, in some sense, the phenomenon of disappearance. As Saturn's period of revolution extends to nearly thirty of our years, during one-half of this period the inhabitants of the earth will behold one side of the ring, and during the other half they will look upon its opposite surface. All the changes from the greatest opening of the ring, when the planet is seen like a magnificent golden ball, engirdled by its ring of golden light, down to the total disappearance of the ring, require about fifteen years. Then the reverse changes occur, and all the phases and transmutations are accomplished in about thirty years, when they are again repeated in the same order.

The disappearance of the ring, which took place in 1848, was watched by the author at the Cincinnati Observatory with the powerful refractor of that institution. A minute fibre of light remained clearly visible even when the edge of the ring was turned directly to the eye of the spectator. The delicacy of this line far exceeds anything ever before witnessed. When compared with the finest spider's web stretched across the field of view, the latter appeared like a cable, so greatly did it surpass in magnitude the filament of light presented in the edge of Saturn's ring. I had the pleasure of witnessing the phenomena so beautifully described by Sir William Herschel, the movement of the satellites along this line of light, "like golden beads on a wire." This is a consequence of the coincidence of the planes of the orbits of these satellites with the plane of the ring; hence, when the ring is seen edgeways, these orbits will, in like manner, be seen as straight lines, coincident with the line under which the ring is seen.

To add to the extraordinary constitution of this wonderful planet, another ring has recently been discovered by Bond, of Cambridge, and by Lassell, of Liverpool, more mysterious, if possible, than those previously known. This ring lies between the planet and the bright ring, and is of a dusky hue, and only discernible in powerful telescopes. Its outline is the same as that of the other rings, with the inner edge of the smaller of which it seems to unite. This extraordinary appendage is so constituted as to reflect but little light, and is sufficiently translucent to permit the body of the planet to be seen through its substance. I have frequently examined this dusky ring with the Cincinnati refractor, and have sometimes been confident that its breadth at the extremities of its longer axis was much greater than that which would be due to an elliptical figure concentric with the bright rings.

Knowing, as we do, the distance of Saturn, it is easy, from the measures of the diameter of his surrounding rings, to compute their absolute dimensions. The exterior diameter of the larger ring is no less than 176,418 miles, and its breadth is 21,146 miles. The exterior diameter of the second ring is 157,690 miles, leaving a chasm between the bright rings of 1,791 miles across. The breadth of the second ring is 34,351 miles, and the interval between the planet and this ring is 19,090. miles. The thickness of the rings is a matter of conjecture, as it is too minute a quantity to be obtained by any means of measurement at present within our reach. Sir John Herschell does not believe it can exceed 250 miles. A single second of arc, at a distance equal to Saturn, subtends nearly 5,000 miles; so that a bright globe of 5,000 miles in diameter, removed to Saturn's distance, would be covered by the smallest *spider's web* stretched across the field of view of the eye-piece of the telescope. In case we admit the rings of Saturn to be 250 miles in thickness, then, when seen edgeways, the filament of light seen reflected from the outer circumference is only one-twentieth part the diameter of the spider's web.

We pass now to an examination of the

SATELLITES OF SATURN.—The largest of these satellites was discovered by Huygens as early as 1665. Four others were discovered some thirty years later by Cassini. Two more were added by Sir William Herschel on the completion and application of his grand reflector in 1789, while an eighth satellite was discovered by two observers, Bond and Lassell, on the same night (Sept. 19th, 1848), the one in Cambridge, United States, the other in Liverpool, England. We have thus, in addition to the anomalous rings which surround Saturn, a scheme of no less than *eight* dependent worlds, all of which revolve about the central orb in elliptical curves, and in periods varying from twenty-two hours to seventy-nine days. If the celestial scenery of Jupiter is rendered magnificent by the splendor of his four moons, what must be the amazing grandeur of the nocturnal sky of Saturn, arched from horizon to horizon by his broad, luminous girdle (on which the shadow of the planet, like the dark hand of a mighty dial, will mark the hours of the night), the changes, phases, eclipses, the occultations of his numerous moons, and the brilliant background of glittering constellations which gem our nocturnal sky, must altogether form a display of celestial splendor of which the human mind can form but a faint conception.

In consequence of the vast distance at which the Saturnian system is removed, and the magnitude and power of the telescope demanded for its examination, we are as yet comparatively ignorant of many facts, which, in the case of Jupiter's moons, have been well determined. It will be remembered that the moon's distance from the earth is about 237,000 miles. Three of Saturn's moons fall far within this limit, and the fourth is but 243,000 miles from its primary. The fifth is 340,000 miles distant; the sixth, 788,000 miles; the seventh (latest discovered), is about 1,000,000 miles distant, while the eighth is removed from Saturn to a distance of nearly 2,300,000 miles.

The nearest of the moons, revolving at a distance of 120,000 miles, circulates round the primary in about twenty-two hours and a half, presenting all the phases exhibited by our moon, in less than a thirtieth part of the time. Its disk, as seen from Saturn, will surpass the moon's disk in the ratio of ten to one. Of the five earliest discovered satellites, two are readily seen with any good telescope. The five may now be seen by many refractors and reflectors of modern construction, while the three smallest satellites are only rendered visible by a few of the most powerful instruments in the world.

We shall here close what we have to present of the structure of the Saturnian system. We have thus terminated the examination of all planetary bodies known to the ancients, and have added to these the new objects revealed by the telescope, inclosed by the circumscribing orbit of Saturn. Within these limits we find all the phenomena known to the master minds to whom we are indebted for the vast extension of the boundaries of human knowledge in the solar system.

Before we pass these old limits, which for so many thousand years were regarded as impassable, we must render an account of the great discoveries, whereby it became possible to achieve the crowning victories of human genius in the planetary regions, and to extend these conquests far beyond the limits of solar influence into regions of space and among revolving orbs, of which the old philosophers had no conception.

CHAPTER IX.

THE LAWS OF MOTION AND GRAVITATION.

The Demands of Formal Astronomy.—Those of Physical Astronomy.—Synopsis of the Discoveries already made.—Questions remaining to be Answered.—Inquiry into Causes.—The Laws of Motion demanded.—Rectilineal Motion.—Falling Bodies.—Law of Descent.—Motion of Projectiles.—Curvilinear Motion.—First Law of Motion.—Second Law of Motion.—Momentum of Moving Bodies.—Motion on an Inclined Plane.—The Centrifugal Force.—Central Attraction.—Gravitation.—Laws of Motion applied to the Planets.—Questions Propounded in Physical Astronomy.—Newton's Order of Investigation.—His Assumed Law of Gravitation.—Outline of his Demonstration.—Its Importance and Consequences.—The Law of Gravitation embraces all the Planets and their Satellites.—Gravitation Resides in every Particle of Matter.

The discoveries thus far made among the revolving worlds dependent on the sun have their origin in a rigorous comparison between the actual phenomena presented in nature and the hypothetical facts derived from an assumed theory. Hipparchus and Ptolemy surpassed their predecessors because, on careful examination, they discovered that the motion of the sun and moon and planets, not being uniform, as had been asserted and believed, they explained this irregularity by the hypothesis of an eccentric position in the central orb, thus enabling them to anticipate all the anomalous movements known in the age in which they lived. Succeeding discoveries, adding to the complexity of the theory of eccentrics and epicycles, drove Copernicus to a new center of motion in the sun, and this hypothesis, united to the old theory of epicycles, was sufficient to harmonize the then known facts of astronomy with the predictions of scientific men.

Increased accuracy of observation, however, soon revealed certain undoubted discrepancies between the absolute places of the heavenly bodies, as given by instrumental observation, and their places as obtained by computation, and after exhausting every possible expedient to restore harmony between observation and computation, finding it impossible, Kepler, as we have seen, was compelled to abandon the circular theory, as Copernicus before him had been forced to relinquish the geocentric hypothesis.

In all this long lapse of many thousands of years the human mind has occupied itself exclusively with the great problem of framing an hypothesis which would embrace all the phenomena as presented in the heavens. It was a question as to *where* was the center of motion, not why it was there; what was the figure of the planetary orbits, not *why* this particular figure existed; *how* the planets deviated from a uniform velocity in their revolution round the sun, not *why* they were accelerated and retarded; *how* the periods of revolution and the mean distances were related, not *why* they were thus related. In short, the facts and not the causes occupied the exclusive attention of the great astronomers, until the science of facts, or *formal astronomy*, had reached its limit, and the mind, having exhausted this field of investigation, was compelled to turn its attention to *causes* or to *physical astronomy*.

Let us review and condense the facts thus far developed by formal astronomy.

The planets revolve about the sun as their common center of motion in orbits whose figure is nearly, if not quite, elliptical.

Their motion is not uniform, but grows swifter as they approach the sun, and loses in velocity after passing their perihelion or nearest point to the sun.

The dimensions of the planetary orbits are not absolutely invariable. There are slight fluctuations not to be overlooked in the periodic times and mean distances.

The positions of the orbits in their own plane are subject to perpetual change, very slow in some of the planets, but comparatively rapid in the moon and in the satellites of other planets.

The inclinations of the planes of the planetary orbits to a fixed plane are also in a state of fluctuation, some angles increasing while others are diminishing.

The lines of nodes, in which the planes of the orbits of the planets intersect the plane of the earth's orbit, are not fixed lines. They are found, on the whole, to retrograde, but are sometimes found to advance in the order of planetary revolution.

The moon exhibits anomalous movements, very marked and well-defined, and which are evidently outside of her elliptic motion, rising above and superior to the general law of her revolution.

The most considerable of these lunar inequalities amounts to 1° 20′ 30″, by which quantity she is alternately in advance and behind her elliptic place in her orbit. This motion was known to the ancients, having been discovered by Ptolemy, and was readily appreciable by the imperfect instruments employed by the Greek astronomers. It is known as the moon's *evection.*

A second inequality, amounting to 1° 4′, called the

moon's *variation*, was discovered by the Arabians, and by them transmitted to posterity, showing the moon's motion accelerated in the quadrants of her orbit *preceding* her conjunctions and oppositions, and retarded in the alternate quadrants.

A third lunar inequality was called the *annual equation*, a name adopted to express the fact that the moon's place in her orbit is for half a year in advance of her elliptic place, and for the other half year behind it.

The line of apsides or longer axis of the lunar orbit performs a complete revolution in the heavens in 3232.57 days.

The line of nodes revolves round the heavens in 6793.39 days.

The vernal equinox is also in motion, sweeping round the heavens in 25.868 years, while the north pole of the heavens revolves round the pole of the ecliptic in the same exact period.

Add to these facts, all discovered by observation and reflection, the grand discovery of Kepler, that the squares of the periods of revolution are precisely proportional to the cubes of the mean distances of the planets, and that this and the other laws of Kepler govern the satellites, and we have a fair exhibition of the great truths of *formal* astronomy, and it is to answer *why* these phenomena exist that *physical* astronomy has been cultivated as a science.

Why does a planet continue to revolve about the sun? In case it approach the sun at all, why not continue that approach until it be precipitated on the surface of the solar orb? *Why* do these revolving bodies describe elliptic orbits, with the sun always in the focus of every orbit? Why are the deviations from elliptic motion what they are? and how comes it that the elliptic elements are in a state of perpetual fluctuation? Why do not the planets fall on the sun, or fly off into space, or stop motionless in their orbits? What holds the earth to the sun, or the moon to the earth, so that they never part company, and unitedly sweep harmoniously round the sun? What bond unites all the satellites to their primaries, and all these primaries to one central orb?

Do all these interrogatories admit of a single answer, or shall we find these phenomena to spring from diverse origins, and due to a variety of causes?

Before it was possible to consider any one of these grand problems it became necessary to reconstruct the old science of mechanics or mechanical philosophy, which was contemporaneous with the Greek astronomy of Ptolemy, and at the time of Kepler and Galileo exerted quite as powerful an influence over the human mind as did the doctrines of Ptolemy and Hipparchus. The philosophy of Aristotle was taught in all the schools, sustained by the immense influence of professional organization, and received with a fullness of confidence and depth of submission which, so far from tolerating doubt, actually prohibited inquiry.

As the planets and their satellites were bodies *in motion*, no advance could be made in the inquiry concerning *causes* until the true nature of motion and the laws by which it was governed could be determined. These laws could only be revealed by accurate thought and observation, and would naturally be independent of the cause producing the motion.

The most obvious example of motion is where a heavy body is dropped vertically from any height, and falls toward the earth. Observation teaches the rectilineal path of such a falling body, as well as its direction, which is toward the center of the earth. It was a matter of experiment to determine whether the velocity was uniform or accelerated, and if accelerated, observation alone could determine the law of acceleration. All this was quite independent of the cause producing the original motion, the rectilineal direction, the acceleration, and the direction toward the earth's center.

Aristotle had laid down the law of falling bodies, and asserted that in case balls of unequal weight were dropped at the same moment from equal elevations the heavier ball would move the swifter, and that the velocity of the two balls would be directly proportional to their respective weights. It was easy by experiment to prove or disprove this statement, and Galileo is said to have given the first powerful blow to the Greek philosophy, by showing experimentally, (by dropping balls of unequal weights from the summit of the leaning tower of Pisa,) that the velocity of the ball, or the time occupied in the fall, was entirely independent of the weight of the ball, the resistance of the atmosphere being taken into account.

By measuring the space passed over by a falling body in equal intervals of time, it became possible to determine the law of descent, and it was thus found that every falling body passes through, say, sixteen feet in the first second of its fall. This is the velocity impressed in the first second of time, and were the body to move on with the velocity thus acquired, it would pass uniformly over thirty-two feet in the next second of time. But it is found that the velocity of every falling body is increasd in every second of time by the same precise velocity acquired in the first second, and thus in case a cannon ball were projected downward at the rate of a thousand feet in one second, it would not only pass over one thousand feet, but the sixteen additional feet acquired by a body falling from a state of rest would be added to the thousand feet due to the impulsive force of the gunpowder.

Again, in projecting a body vertically upward, it was discovered by experiment that at equal elevations in the ascent and descent the velocities were identical, and thus whatever might be the cause retarding the ascending body, or accelerating the descending one, that cause was found to exert its force with a constant energy.

Galileo was the first fully to develop the facts above stated, which facts manifestly began to couple motion and velocity with some mysterious cause of acceleration and retardation.

The rectilineal motion of falling bodies naturally led to the inquiry as to the line in which a body, receiving a single impulse, would move, if entirely free from the influence of extraneous forces. A body shot from a gun horizontally, at the commencement of its motion seemed to move in a right line, but a more rigid examination showed that (the air as a resisting medium out of consideration) the bullet commenced to fall at the moment of its flight, and actually did fall in one second of time through a vertical height precisely equal to the space it would have fallen through if dropped from the muzzle of the gun. Here, then, was a deviation from a rectilineal path accounted for by admitting a constant deflecting force precisely equal to that force, wherever it may be lodged, or whatever it may be, which produces and accelerates the velocity of falling bodies.

As the right line is the most perfect of all lines, and

as uniform motion in a right line is the simplest of all motions, Galileo conceived the idea that in case a body receive a single impulse giving a velocity of any rate per second, that the body thus set in motion will move off with uniform velocity in a straight line, holding the direction in which the impulse is applied, and will thus continue moving for ever, unless some force or power be exerted to change the direction or to destroy the velocity.

This conception or hypothesis could not be proved *directly* from experiment. A ball perfectly hard, round and smooth, shot on a level surface like ice, would preserve its rectilineal path and its initial velocity much longer than if opposed by irregularities of surface and other resisting causes; and thus it became manifest that as the resistances to motion were diminished, there was a nearer positive and experimental approach to the verification of the principle laid down by Galileo, till finally it became a settled principle, and was at length dignified as the *first great law of motion.*

Previous to the discovery of this law the mind had never been able to conceive the idea that motion could continue after the cause producing it had ceased to act, and yet there is no motion produced by human contrivance, such as the motion of a stone from a sling, or a ball from a cannon, in which it is not manifest that the force producing the motion ceases its action after the first impulse. The sling cannot pursue the stone once liberated, nor the powder with its expansive power follow the ball once released from the gun; and thus it is clear that motion, once generated, survives for a longer or shorter time the direct action of the impulsive force.

So much for rectilineal motion. We are indebted to

Galileo again for the second law of motion, or the law by which we pass from rectilineal to curvilinear motion. A ball projected horizontally, as we have seen, soon commenced to fall away from the straight line in which its motion commenced. Galileo proved that under the united effect of the projectile force, and the force which caused it to fall toward the earth, the ball would describe a regular curve, called a *parabola*, which is nothing more than an ellipse, whose major axis is infinitely long. This curve may be seen in the form of the jet, when water or any other fluid spouts from an orifice near the bottom of a cylindrical vessel.

The Florentine philosopher in pursuing this subject finally came to generalize the principle involved, and discovered that if a body in motion at any given rate receive an impulse, whose line of direction forms any angle with the line of direction of the moving body, it will immediately take up a new line of direction, according to a law which may be thus announced:—

If two sides of a square or rectangle represent the intensities and directions of two impulsive forces acting at the same instant, on a body at the angle formed by these sides, then the body which, at the end of one second, would have been found, under the impulse of either force, at the extremity of the side representing the force, will neither follow the one side of the rectangle nor the other, but will take the direction of the diagonal, and at the end of one second will be found at the *extremity of that diagonal*, as may be more readily comprehended from the figures below.

Let F be any impulsive force, such, that acting on the material point P, in the direction P B, would project it to B in one second of time, and F′ be an impulsive force which, acting on the same material point P, in the direction P A, would project it to A in one second; then, in case the two forces operate at the same moment on the material point P, at the end of one second it will neither be found at B nor at A, but will be found at C, the extremity of the diagonal of the parallelogram.

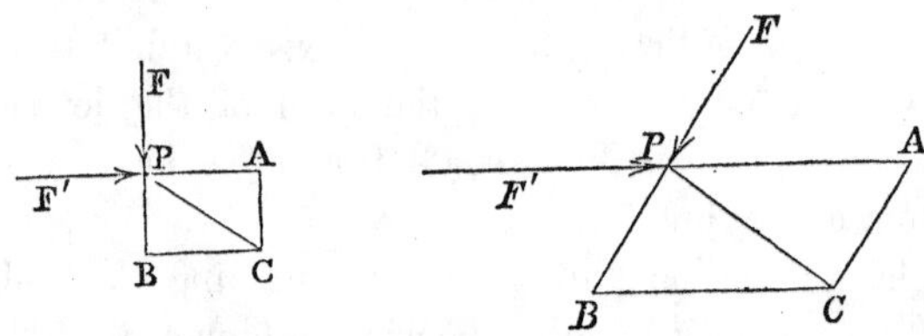

This principle is known as the *second law* of motion, and is also known as the *parallelogram of forces*.

In these investigations no account has been taken of the weight or mass of the body moved. It was clearly perceived that the force exerted by a body at rest pressing upon any support was precisely proportioned to its weight, and hence a ball weighing ten pounds would bend a spring through ten times the space due to a ball weighing but one pound. Aristotle knowing this truth, and believing that this pressure downward was the moving force, when a body fell freely, asserted the principle that a body would fall with a velocity proportioned to its weight, which, as we have said, was disproved by Galileo in his celebrated experiment at the leaning tower of Pisa.

It was manifest, however, that when two balls of unequal weight fell from the same elevation, although they struck the ground at the same moment, or fell with equal velocities, the effect of the blow struck by the heavy body was very different from that produced by the lighter body. Indeed, it was easy, by experiment, to prove that the effect was precisely proportioned to the weight of the falling body, and that a body of ten pounds weight would strike a blow ten times more powerful than a one pound weight after falling through equal heights. It was thus seen that to estimate the effect of a blow struck by a moving body we must take into account not only the velocity but also the weight or mass of the body. The same body, with double velocity, doubling its effect, in short, the mass multiplied by the velocity, now called the *momentum*, became the true representative of the effect produced by the blow struck by a moving body.

Having reached clear ideas and true laws on these important subjects, Galileo gave his attention to the circumstances of motion on an inclined plane. By experiment he demonstrated that, if the same body roll down planes of the same vertical height, but with different inclinations, the velocity acquired on reaching the foot of any one of these planes will be independent of the inclination, and will always be equal to the velocity due to the vertical height of the inclined plane. This discovery presented the principle of the *third* and last law of motion, and, after much discussion, came to be adopted as a fundamental truth in mechanical science. These laws of motion were the result of clear reasoning, based upon accurately conducted experiment, and were quite independent of the actual causes producing motion.

So soon as the knowledge of the second law of motion was attained, whereby it became demonstrable that a body set in motion by a single impulse, and then operated on by a constant power, would describe a curve, it seems strange to us, surrounded as we now are by the full illumination of a true science, that this principle was not directly applied to account for the motions of revolution of the celestial orbs.

Kepler, whose fertile genius, ever active and untiring, sought the cause of planetary motion, being ignorant of the laws of motion, felt that he must discover and reveal some constantly active power operating in the direction of the planetary motion so as to keep up the velocity, believing that without some such ever-active force the planets must of necessity stop. The successors of Kepler and of Galileo, for fifty years, or during the first half of the seventeenth century, felt strongly the necessity of a physical theory of the planetary motions, without attaining to anything clear or satisfactory.

That all heavy bodies were in some way attracted by the earth, and that the center of attraction was in the earth's center, was manifest from the fact that every body falling freely, sought the earth's center. But how a central force, lodged in the earth or in the sun, could operate to keep up a motion of revolution round that center, in distant bodies, was the inexplicable mystery. The ancients had already remarked that when a stone in a sling is whirled rapidly round the head of the slinger, a force is developed which powerfully stretches the string by which the stone is held. This force was called the *centrifugal force*, and it finally came to be accepted that, in all revolving bodies, this tendency to fly from the center must be generated, and hence in the planets and their satellites a like tendency must exist. Reasoning, then, upon the two great facts, that all bodies gravitate to the earth, and from analogy all bodies equally gravitate to the sun, and that all revolving bodies, by the action of the centrifugal force generated in their revolution, are disposed to fly from the center of motion, Borrelli, of Florence, in 1666, seems to have been the first to conceive the idea that in the planets and their satellites these two forces might mutually destroy or counterbalance each other, and leave the planet in a state of dynamical equilibrium to pursue its journey round the sun.

Here we find the germ of the grand theory which at the present day embraces within its grasp the entire physical universe of God. It was, however, but the germ. Borrelli did not pretend to demonstrate the truth of his suggestion. To accomplish this, it became necessary to demonstrate the law of the centrifugal force and the law of gravitation, and then to show that the first of these forces, as developed by the velocity of revolution of any planet, was precisely equal to the force of gravitation exerted by the sun at the distance to which the planet was removed.

The law governing the development of the centrifugal force could be investigated experimentally. A cord sufficiently strong to hold a body suspended from a fixed point was not strong enough to hold the same body when made to revolve about the point, and, as the velocity of revolution was increased, the strength of the cord had to be increased. But it was soon found that, with double the velocity a cord twice as strong would not retain the revolving body. The centrifugal force increased, therefore, in a higher ratio than the simple velocity. By further experiment it was discovered that when the velocity of the revolving body was *doubled*, the cord holding it must be *quadrupled*, and when the velocity was tripled the cord must be made nine-fold stronger, and hence it became finally a fixed principle *that the centrifugal force in every revolving body increased with the square of the velocity.* It remained yet to ascertain in what way this force was affected by the distance of the revolving body from the center of motion. This was accomplished experimentally, and the complete law regulating the development of the centrifugal force in all revolving bodies having been determined, this force was found to increase as the square of the velocity and to decrease directly as the distance from the center of motion increased.

With the knowledge of this important law, we can return to the consideration of the planetary revolutions. That these bodies were urged directly from the sun by the action of the centrifugal force generated by their velocity of revolution could not be doubted, and to counterbalance this tendency to fly from the center of motion some force precisely equal and opposite must exist. This force was called the *gravitating force*, or *force of gravity*, and the law regulating its intensity remained to be discovered.

Kepler had not failed to conjecture the existence of some such central force lodged in the sun and in the earth. He even went further, and conceived the same force of attraction to exist in the moon, and finding the tides of the ocean to be swayed by that distant orb, he conceived that the same energy which manifested itself in a heaving up of the ocean wave, must exert itself with equal power on the solid mass of the earth. These, however, were mere speculations with Kepler, and even, in case he had seriously undertaken to prosecute the research, the ignorance of the true laws of motion then existing would have rendered any success impossible. From the days of Kepler to those of Newton this great problem constantly occupied the thoughts of the most eminent philosophers: Was the gravitating force whereby bodies fell to the earth's surface a constant or variable force? Was this force operative in the distant regions of space? Did its power extend to the moon? and was it there precisely what it should be in order to counterbalance the energy of the centrifugal force? Did this same gravitating power dwell in the sun and other planets, as well as in the earth? Did the sun's gravity extend to each of the planets, and exert at these different distances an energy equal and opposite to the existing centrifugal force due to the planet's distance and velocity? In short, was there a force or energy dwelling in every particle of ponderable matter whereby every existing particle attracted to itself every other existing particle, with an energy proportioned in some way to their weights and to the distances by which they were separated? Could such a force, lodged centrally in the sun, and operating by any law, convert the rectilinear motion of a body darted into space by a single impulse into elliptical motion, and at the same time, at every point in the elliptical orbit, precisely counterpoise the centrifugal force due to the planet's distance and velocity? Could the same force, governed by the same laws and lodged in the primary planets, control the movements of their satellites? These were the grand inquiries which engrossed the attention of the generation of philosophers which flourished from the time of Kepler and Galileo up to the era rendered immortal by the grand discovery of the law of universal gravitation by Newton.

NEWTON'S DISCOVERY OF THE LAW OF GRAVITATION.—We are now prepared to consider the train of reasoning employed by the English philosopher in his researches for the law of gravitation. Many astronomers before Newton had conjectured that the force exerted by the sun on the planets, and by the primaries on their satellites, *decreased* as the *square of the distance increased*, or followed the law of the inverse ratio of the square of the distance. This was inferred from the consideration of fact, that this attractive energy, called gravity, was lodged in the center of the sun, and issued from that center in all possible directions, like light emanating from a luminous point. As the distance increased from the center the force would become less intense, and might follow the law of the decrease in the intensity of light, which was well known by experiment to be the inverse ratio of the square of the distance. It was one thing to conjecture this to be the law regulating the force of gravity, but quite a different thing to demonstrate the truth of such a conjecture.

The investigation as pursued by Newton, and the discoveries made by that distinguished philosopher, followed progressively in a series of distinct propositions, the demonstrations of which were reached at different periods.

First, Newton demonstrated that, assuming the third law of Kepler as a fact derived from observation, as a consequence of this fact, (combined with the law of the centrifugal force,) the gravitation of the planets to the sun must diminish in the inverse ratio of the square of their respective distances. This demonstration was accomplished by a train of mathematical reasoning, of which we will not stop to give any account at present. It was based, however, on the assumption that the planetary orbits were circles, and hence did not meet the case of nature.

The *second* step was to prove that in case a planet revolved in an elliptical orbit, that at every point of its revolution the force exerted on it by the sun, or its gravitation to the solar orb, was always in the inverse ratio of the square of its distance. This was equivalent to proving that if a body in space, free to move, received a single impulse, and at the same moment was attracted to a fixed center by a force which diminished as the square of the distance at which it operated increased, such a body, thus deflected from its rectilineal path, would describe an ellipse, in whose focus the center of attraction would be located.

The *third* step in this extraordinary investigation was to demonstrate that, this gravitating power lodged in the sun, and controlling the planetary movements, was identical with that force exerted by the earth over every falling body, and extending itself to the moon, decreasing in intensity in proportion as the square of the distance increased, and thus opposing itself as a precise equipoise at every moment to the effect of the centrifugal force generated by the motion of this revolving satellite.

The *fourth* step required the philosopher to demonstrate that not only did the planets gravitate to the sun, and the satellites to their primaries, but that each and every one of these bodies, sun, planet and satellite, gravitated to the other, and that each attracted the other by a force which varied in the inverse ratio of the square of the distance. But here it was found that another matter had to be taken into account. The energy of gravitation did not depend alone on *distance.* The power exerted by the sun on the planet Jupiter was vastly greater than that exerted by Saturn, though Jupiter was nearly equidistant from these two bodies when in conjunction with Saturn. Newton proved that the power of gravitation lodged in any body depended on the *mass* or *weight* of the body, and hence if the sun weighed a thousand or ten thousand times as much as a planet, its energy at equal distances would, by so much, exceed that put forth by the smaller orb.

The *fifth* and final step in this sublime, intellectual ascent to the grand law of the physical universe, required the philosopher to prove that the force, power, or energy, now called gravitation, lodged in the sun, planets and satellites, pervaded equally every constituent particle of each of these bodies, and did not dwell alone in the mathematical center of the sun or planet. In short, it was required to show that every ponderable particle of matter in the whole universe possessed and exerted this power of attraction *in the direct proportion of its mass, and in the inverse ratio of the square of the distance at which its energy was manifested.*

In case these propositions could be clearly and satisfactorily demonstrated, an instant and absolute revolution must commence in the whole science of astronomy, and the business of future ages would be nothing but the verification of this one grand controlling law, in its application to the phenomena presented in the movements not only of the sun's satellites and their attendants, but in those grander schemes of allied orbs revealed by telescopic power in the unfathomable regions of the sideral heavens.

We shall now exhibit an outline of the demonstration accomplished by Newton to prove that the law of universal gravitation, as above announced, was the exact law according to which the earth exerted its attractive power on the moon, and held this globe steady in its orbit.

The intensity of any force, as we have seen, is measured by the velocity it is capable of producing in a heavy particle in any unit of time, as one second. The earth's gravity at the surface is measured then by the space through which a body falls in a second of time, which space (as experiment demonstrates) is about sixteen feet. In case it were possible to measure with absolute precision the space through which a body falls at the level of the sea and then at the summit of a mountain (if there were any such) 4,000 miles high, it would be easy to verify the truth of the assumed law by actual experiment; but no mountain exists on the earth's surface whose height is comparable with the length of the earth's radius, and as it is absolutely impossible to ascend vertically above the earth to any considerable height, Newton soon saw that no means existed on the surface of the earth whereby the truth or falsehood of his assumed hypothesis might be ascertained. In this dilemma he conceived the idea that the moon might be employed in the experiment, not by arresting her motion and dropping her literally to the earth, but by considering the earth's attractive power as the cause of her deflection from a rectilineal movement. In one sense the moon is perpetually falling to the earth, as may be readily comprehended from an examination of the figure below :—

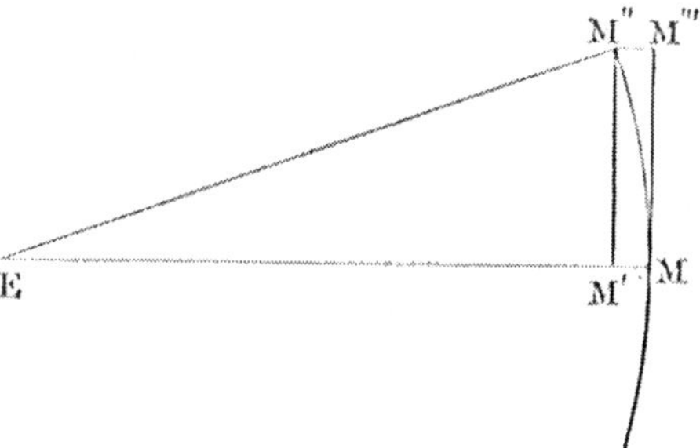

Let E represent the earth's center, M a point of the moon's orbit, in which she is at rest with no force whatever operating on her. Now let an impulse be applied to the moon, in the direction M M‴, tangent to the orbit, or perpendicular to the line M E, and with such intensity that at the end of one second of time the moon will be found at M‴. Return the moon to M, and conceiving her to drop toward the earth, under the power of the earth's attraction, let us suppose that she passes over the distance M M′ in one second. In case the moon be brought back again to M, and the impulse be now applied, and at the instant the moon darts off along the straight line M M‴, she is seized by the earth's attractive power, and, bending at once under these conjoined influences, she commences her elliptical orbit, and at the end of one second is found at M″, passing over a sort of curvilinear diagonal of the parallelogram formed on the two sides M M‴ and M M′. Now, it is manifest that the line M‴ M″ is equal to M M′, that is, that the amount by which the moon is deflected from a right line is precisely the amount by which she falls to the earth in one second of time. The problem then resolved itself into a computation of the line M M′, or the distance through which the moon ought to fall in one second, in case the assumed law of gravition be true, and the exact measurement instrumentally of the distance M‴ M″, the space through which the moon did fall in one second. An exact equality between these two quantities would establish the law of the decrease of the earth's power of attraction to be in the inverse ratio of the square of the distance.

It will be seen that to compute how far a body would fall in one second, when removed to the moon's distance, in case the earth's gravity be diminished as the square of the distance increases, is a matter involving no difficulty or uncertainty whatever, in case we know what the moon's distance is. In like manner, to obtain the space through which the moon actually falls to the earth in one second or minute of time, knowing her distance, admitting her orbit to be circular, and assuming that we know her periodic time, is a problem of easy solution.

The chief difficulty lay in accomplishing an accurate determination of the moon's distance from the earth, a matter which could not be determined without an accurate knowledge of the earth's diameter or radius, as we have already seen.

When Newton commenced his investigation the measures which had been executed of an arc on the meridian, whereby the entire circumference of the earth might be obtained and its diameter computed, were comparatively imperfect, yielding only an approximate value of the earth's radius. As this quantity was the unit employed in the measure of the moon's distance, any error in its value would be repeated some sixty times in the value of the moon's distance, and as the gravity of the earth was assumed to decrease as the *square* of the distance increased, we perceive that any error in the radius of the earth would operate fatally on the solution of this problem, involving the fate of the most comprehensive and far-reaching hypothesis ever conceived by the human mind.

Unfortunately for Newton, the value of the earth's diameter, employed in his first computations, was in error, and in executing the computation the values of the space through which the moon *ought* to fall, and the space through which she *did* fall, were discrepant by an amount equal to the sixth part of the entire quantity. So great a disagreement was fatal to the theory in the truth-loving and exact mind of Newton, and for many years he abandoned all hope of demonstrating the truth of his favorite hypothesis. Still his mind was in some way powerfully impressed with the conviction that he had divined the true law of nature, and he returned again and again to his computations in the hope of removing the discrepancy by detecting some numerical error. It was impossible, however, to find an error where none existed, and for a time the great philosopher abandoned all hope of accomplishing this, the grandest of all the efforts of his own sublime genius. Such was the condition of this investigation when a new determination of the value of the earth's diameter was accomplished in France, by the measurement of an arc of the terrestrial meridian. Having obtained this new value of the earth's diameter, Newton resumed once more the consideration of the problem which had so long occupied his thoughts. The new value was substituted for the old—the moon's distance being now accurately known—the space through which a body would fall in a unit of time, under the power of gravitation, when removed to this new distance, was rapidly computed. In like manner the distance through which the moon must actually fall was also obtained by using the new value of the earth's diameter.

It would be impossible to form any just idea of the intense emotions which must have agitated the mind of the English philosopher while engaged in bringing these last computations to a close. Upon a comparison of the results now reached there hung consequences of incalculable value. No less than nineteen years of earnest study, of profound thought, and of the most laborious investigation, had already been exhausted on this grand problem, and now within a few minutes the fate of the theory and the fame of the astronomer were to be for ever fixed. No wonder, then, that we are told that even the giant intellect of Newton reeled and staggered under the tremendous excitement of the moment; and seeing that the figures were so shaping themselves as inevitably to destroy the discrepancy which had so long existed, overcome by his emotions, Newton was compelled to ask the assistance of a friend to finish the numerical computation, and when completed it was found that the space through which the moon *did* fall in a unit of time was identical with the space through which she *ought* to fall, in case her movements were controlled by a power lodged in the earth's centre, and decreasing in energy as the square of the distance at which it operated was increased.

Here was presented the first positive proof of the prevalence of that universal law of mutual attraction which energizes every particle of ponderable matter existing in the universe. The earth's power of attraction was thus shown to exert itself according to a fixed law, in deflecting the moon from the rectilineal path it would otherwise have followed, converting its motion into one of revolution, giving to its orbit the elliptical form, and maintaining at every point of its revolution the most exact and perfect equilibrium.

It may, perhaps, seem extraordinary that so much consequence should have been attached by Newton to the successful demonstration of this particular problem. If he had already shown that the sun's attraction upon the planets followed the law of the inverse ratio of the square of the distance, and that the same law prevailed in the attraction of the sun upon any one planet at different points of its orbit, why regard as a matter of such high value the demonstration of the fact that the earth's attraction upon the moon was governed by the same identical law. The answer, I think, may be readily given.

The great problem was this: Does one law reign supreme over all the ponderable masses of the physical universe, or are there many subordinate laws holding their sway in the diverse systems and bodies which are revealed by sight? Might it not be that the sun would attract the planets according to one law, while the planets might attract their satellites according to a different law? By demonstrating that the earth controlled the moon by the same precise power whereby the sun controlled the planets, it was demonstrated that the ponderable matter of the earth was identical in character with the ponderable matter of the sun, and from this it followed that as the earth was one of the planets controlled by the power of gravitation of the sun, so likewise the other planets which were controlled by the same power must be composed of ponderable matter, governed by the same laws which reign in the sun and earth.

We perceive, then, that this demonstration, executed by Newton, in which he proves that the earth's attraction controlled the moon, deserves the high rank which he has assigned it, for it is nothing less, when conjoined with his previous demonstrations, than proving that every globe which shines in space, planet and satellite, and sun, are but parts of one mighty system linked together by indissoluble bonds, forming one grand scheme, in which each exerts its influence upon the other, the whole controlled by one supreme and all-pervading law.

It only remained now to extend by demonstration the empire of the law of universal gravitation over each particle of matter composing the several worlds. This was a problem of no ordinary difficulty; for Newton soon discovered that in case a mass of ponderable matter were fashioned into the shape of a sphere, that for all the purposes of computation it would be safe to consider the entire globe as concentrated in one single point at the centre. Observation taught that all the planets, as well as their satellites, were bodies of globular form, and hence in applying the law of universal gravitation to the study and computation of their movements, the same results would be obtained by admitting the whole force of attraction belonging to these bodies to be concentrated in their central points, or to be distributed among the different particles composing the globe. To show, then, that gravity resided in every particle composing a globe, and not in its central point, was an impossible thing, so far as the distant worlds were concerned. In the world which we inhabit, however, and where we can study its individual portions, and where we can penetrate to certain depths toward its center, it may not be impossible to learn whether the power of gravitation dwells in every ponderable atom which goes to make up the entire earth, or whether it is concentrated in the central point alone.

There are several methods which may be employed to ascertain whether there be any power of attraction in separate portions of the earth or in the crust of the earth. The effect of a high mountain on the direction of the plumb-line, (which at the level of the sea holds a direction perpendicular to the surface,) in causing it to deviate from this direction, may be measured with sufficient accuracy to demonstrate the power of attraction existing in the mountain. Such an experiment, however, could not be employed to demonstrate that the law of universal gravitation prevailed among the particles composing the mountain, it would only show that there was a power of attraction exerted by the mountain, and in case we knew the exact amount of deviation of the plumb-line from the vertical, and the magnitude of the mountain, as well as the law according to which its attractive power was exerted, we could then obtain the quantity of matter contained in the mountain mass.

A second method may be employed to ascertain whether the whole power of gravitation is lodged in the center of the earth or is distributed among all its constituent particles. If it were possible to penetrate toward the earth's center, a thousand miles below the surface, and there drop a heavy body, and measure the space through which it falls in a unit of time, if this measured space should be identical with that over which the body ought to fall, on the supposition that its velocity depended simply on its distance from the center, such an experiment would demonstrate that the earth's gravitating force resided in the central point alone; this experiment cannot be performed in the exact manner announced, but it can be, and has been substantially performed, with very great delicacy, in the following manner:

It is found that a pendulum of given length will vibrate seconds at the equator of the earth. If this pendulum be removed nearer to the earth's center by carrying it toward the poles, the power of gravitation producing its vibration thereby growing more intense, the pendulum will vibrate more than sixty times in a minute, and thus the number of vibrations in a given time becomes a very exact means of measuring the distance of any point on the earth's surface from its center. These experiments, however, are performed upon the earth's surface. If, instead of removing the pendulum from the equator toward the poles, and thereby reducing its distance from the earth's center, this distance were reduced by the same amount by transporting the pendulum vertically downward into a deep mine—if distance alone from the center be the cause affecting the time of the vibration of the pendulum—then the number of vibrations in a unit of time will be the same in the mine as at a point on the exterior equidistant from the earth's center. When this experiment comes to be performed it is found that there is a great difference between the number of vibrations in the interior, when compared with the number of vibrations at the exterior, at equal distances from the center, in any unit of time, say a mean solar day, clearly demonstrating that the matter of the earth, lying above the pendulum located in the mine, produces a very sensible and powerful effect upon the number of its vibrations.

Here, again, we find it impossible, from this experiment, to determine the exact law which regulates the attractive power of the individual particles composing the earth, but we do demonstrate the fact that the earth's gravity is not concentrated at its center, but dwells, according to some law, in all the atoms which compose its mass; and this law, we shall prove hereafter, is none other than the great law of universal gravitation.

It is impossible to form a just idea of the vast importance which attaches to the grand discovery of Newton. It worked out, instantly and absolutely, a complete revolution in the whole science of astronomy. Previous to the discovery of the law of universal gravitation all the observations upon the stars and planets, which had been accumulating for so many centuries, could only be regarded as so many isolated facts, having no specific relation the one to the other. The planets were independent orbs, moving through space in orbits peculiar to themselves, and only united by the single fact that the sun constituted the common center of revolution. The discovery made by Newton converted this scheme of isolated worlds into a grand mechanical system, wherein each orb was dependent upon every other, each satellite affecting every other, and the whole complex scheme gravitating to the common center, which exerted a predominant power over each and every one of these revolving worlds.

Those eccentric bodies which we denominate *comets*, whose abrupt appearance in the heavens with their glowing trains of light, whose rapid movements and sudden disappearance have excited such a deep interest in all ages of the world, were found not to be exempt, as we shall hereafter show, from the empire of gravitation.

CHAPTER X.

THE LAWS OF MOTION AND GRAVITATION APPLIED TO A SYSTEM OF THREE REVOLVING BODIES.

A SYSTEM OF TWO BODIES.—QUANTITIES REQUIRED IN ITS INVESTIGATION.—FIVE IN NUMBER.—SUN AND EARTH.—SUN, EARTH AND MOON, AS SYSTEMS OF THREE BODIES.—THE SUN SUPPOSED STATIONARY.—CHANGED FIGURE OF THE MOON'S ORBIT.—SUN REVOLVING CHANGES THE POSITION OF THE MOON'S ORBIT.—SOLAR ORBIT ELLIPTICAL.—EFFECTS RESULTING FROM THE INCLINATION OF THE MOON'S ORBIT.—MOON'S MOTION ABOVE AND BELOW THE PLANE OF THE ECLIPTIC.—REVOLUTION OF THE LINE OF NODES.—SUN, EARTH AND PLANET, AS THE THREE BODIES.—PERTURBATIONS DESTROY THE RIGOR OF KEPLER'S LAWS.—COMPLEXITY THUS INTRODUCED.—INFINITESIMAL ANALYSIS.—DIFFERENCE BETWEEN GEOMETRICAL AND ANALYTICAL REASONING.

WE shall now present, as clearly as we can, without the aid of mathematical reasoning, the application of the laws of motion and gravitation to the circumstances arising in a system of three bodies mutually affecting each other. We will commence even with a simpler case, and suppose a solitary planet to exist, subjected to the attractive power of one sun, and that these are the only bodies in the universe. Let us consider what quantities are demanded to render it possible for the mathematician to take account of the circumstances of motion which will belong to this solitary world.

First of all, it is evident that the *quantity of matter* contained in the sun, or its *exact weight*, must be known, for the energy or power of the sun varies directly as its mass, and two suns, so related that the weight of one is tenfold greater than that of the other, the heavier one will exert a power of attraction tenfold greater than the lighter one.

In the *second* place, we must know the distance of the planet from the sun, for the power of the sun's attraction decreases as the square of the distance at which it operates increases; so that if at a distance of unity it exerts an attractive force which we may call *one*, at a distance *two* this force will be diminished to one-fourth; at a distance *three* to one-ninth; at a distance *four* to one-sixteenth; at a distance *ten* to the one hundredth part of its first value.

In the *third* place, the *mass* or *weight* of the planet must be known; for not only does the sun attract its planet, but in turn the planet attracts the sun, and the intensity of this attraction, which affects the motion of the planet as well as that of the sun, depends exclusively upon the mass or weight of the planet.

In the *fourth* place, we must know the *intensity of the impulsive force* which is employed to start the planet in its orbit, for upon the intensity of this force will the initial velocity of the planet depend, and we see readily that the form of the orbit as to curvature will depend upon the initial velocity. The greater this velocity the more nearly will the curvature of the orbit coincide with the straight line in which the planet would have moved in case it had been operated upon by the impulsive force alone.

In the *fifth* place, before we can completely master the circumstances of motion to the planet, we must know the *direction* in which the impulse is applied, for upon this direction it is manifest that the figure of the orbit will depend. If the inpulsive force be applied in a direction passing through the sun's centre, and toward the sun, it is clear that the planet will simply fall to the sun in a straight line. If it met with no resistance it would pass through and beyond the sun's center until its velocity would be entirely overcome by the attraction of gravitation, when it would stop, fall again to the sun, and thus vibrate for ever in a right line. In case the direction of the impulse is oblique to the line joining the planet and the sun, (the angle falling within certain limits of value,) then the planet will describe an elliptical figure in its revolution around the sun, and will return precisely to the point of departure to repeat the same identical curve, with the same velocities precisely at each of the points of its orbit, in the same exact order for ever. In examining the peculiarities which distinguish the movements of this revolving body, we shall find as a necessary consequence of the laws under which it moves that its motion must be slowest at that point of its orbit where it is furthest from the sun. Leaving this point as it approaches the sun, its velocity must rapidly increase, and will reach its maximum at the perihelion of its orbit, where, being nearest to the sun, it will move with its swiftest velocity. Receding now from the center of attraction it will lose its velocity by the same degrees with which it was augmented, and will again pass its aphelion with its slowest velocity. Thus we perceive that the movements of a single planet revolving about the only sun in existence are marked with great simplicity; and in case the mathematician knows precisely the five quantities already named, viz: *the sun's mass, the planet's distance, the planet's mass, the intensity of the impulsive force, and the direction of this force*, it is not at all difficult to determine all the circumstances of motion of the planet, and to predict its place in its orbit with absolute precision at the end of ten thousand revolutions.

We will not at present attempt to show how these five quantities may be obtained. These determinations belong to the department of instrumental astronomy, a subject which will be treated after closing what we have to say on the application of the law of gravitation to the movements of a system of three bodies.

In case the planets had been formed of a material such as to be attracted by the sun, but not to attract each other, and if the satellites had been composed of a material such as to be attracted by their primaries only, then the elements of the orbits of all these revolving bodies would have remained for ever absolutely invariable. So soon, then, as accurate observation should have furnished the five quantities required in determining the circumstances of motion in any revolving body, mathematical computation would have fitted an invariable orbit to each one of these bodies, and would have furnished by calculation the exact place of each one of these bodies in all coming time. The whole system would have been one of perfect equilibrium, and although complexity would have presented itself apparently in the interlacing revolutions of these revolving worlds, yet absolute simplicity, combined with short periodical changes, would have restored each one of these bodies to the exact position occupied when first launched in its orbit.

This, however, is not the case of nature. The sun not only attracts the planets, but also attracts their satellites. The primary planets not only attract their satellites, but attract each other; and thus not a single body exists in the whole universe which is not dependent upon every other.

We have already seen that in case the sun with one planet were the only objects in existence, that having traced the planet in one single revolution round the sun, the variations of motion thus developed would be repeated without the slightest change in any succeeding revolution for all coming time.

Suppose this solitary planet to be the earth, and that from a knowledge of the weight of the sun, the distance of the earth from the sun, the weight of the earth, the intensity of the impulsive force, and the direction in which that force is applied to start the earth in its orbit, we determine the elements of its orbit. The form of this orbit, its magnitude and position in space will remain absolutely invariable, and the changes of motion in the first revolution will be repeated exactly in all succeeding revolutions. Let us now add to our system of two bodies a third body, as the moon. In case the sun had no existence, or was removed to an infinite distance, then the circumstances of motion in the moon, once determined, would remain absolutely invariable, but the moment we unite the three bodies, the sun, earth and moon, into a system of three orbs, mutually dependent upon each other, the perfection and simplicity which marks a system of two bodies is for ever destroyed, and modifications are at once introduced into the motion of the earth revolving around the sun, and also into that of the moon revolving around the earth, of an exceedingly complex and difficult character, and requiring the highest developments of mathematical analysis to grapple successfully with this great *problem of the three bodies*.

The solution of this problem has never been positively accomplished, but approximations of wonderful delicacy have been reached by the successors of Newton, so that for all practical purposes in astronomy this approximate solution may be fairly regarded as absolute.

As the plan laid down in this work does not admit the use of any but the simplest mathematical elements, we shall only trace out, in general terms, the consequences which must follow from the introduction of a third body into a system of two revolving orbs, and, for the purpose of fixing our ideas, we will suppose the earth and moon to be our two bodies. The moon's orbit, in magnitude, and figure, and position, is supposed to be known; her period of revolution and the circumstances of her motion in her orbit are also supposed to be accurately determined. The earth being fixed in position, and the moon performing her revolution under the laws of motion and gravitation, let us now add to this simple system a third body, the sun; and to render our investigation as simple as possible, we will adopt the hypothesis that the earth continues at rest, but that a new force, namely, the sun's attraction, now commences to exert its influences upon the moon. In order still further to simplify the case, let us suppose the sun's center to be situated in the prolongation of the longer axis of the moon's orbit, and that the moon, in passing through her aphelion, will cross the line joining the centers of the earth and sun. Under this configuration it is clearly manifest that the figure of the moon's orbit will be changed, because the attractive power of the sun will certainly increase the distance to which the moon travels *from* the earth, for the velocity with which the moon moves away from her perihelion point will be reinforced by the attractive power of the sun, and thus her aphelion distance will be increased. By the same reasoning, it will appear that her perihelion distance will be somewhat diminished, and thus the longer axis will be increased in length, and the period of revolution of the moon will, in like manner, be increased.

These changes having been once accomplished, and the moon having taken up her new orbit under the action of the new forces, so long as these forces remain constant, that is, so long as the sun remains fixed in position, the new orbit will remain as invariable as did the old before the introduction of the sun. All the changes accomplished by the sun's power, whereby the new orbit is made to differ from the old, are called, in astronomy, *perturbations*, and the sun is called the *disturbing* body.

Let us now suppose the sun, retaining its distance from the earth, to start from its position on the prolongation of the longer axis of the moon's orbit, and to commence a revolution around the earth in a circular orbit, lying in the plane of the moon's orbit. A little reflection will show us that the moment the disturbing body commences to move, the direction of its attractive power upon the revolving moon will begin to change; a new set of disturbances will now commence, not affecting the *new figure* of the moon's orbit, but changing the position of the principal lines of the orbit in its own plane; for it is clearly manifest that the strongest power will be exerted to draw the moon away from the earth on the line joining the centers of the earth and sun; and hence the aphelion point of the moon's orbit will necessarily try to follow this moving line. The subject will be made plainer by an examination of the figure below, in which E represents the earth in the focus of the moon's orbit, S the place of the sun on the prolongation of P M the longer axis, P the perigee, and M the apogee of the moon's orbit. In case the sun be removed to S′, and there remain stationary, it is manifest that each time the moon crosses the line E S′ it will be subjected to the most powerful influence to draw it away from the earth at E; and in case the sun remain station-

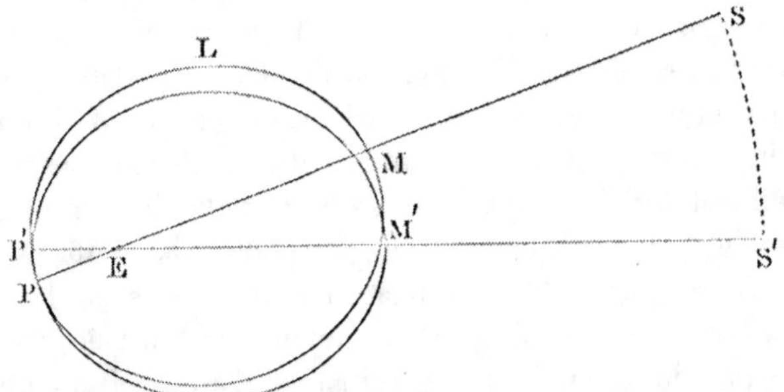

ary for a sufficiently long time at each of the moon's revolutions, the point M will approach M′, and finally it will actually fall on M′, where it will remain fixed, so long as the sun is stationary.

In case, however, the sun again advances in the same direction, the apogee of the moon will again advance, and should the sun, by successive steps, slowly perform an entire revolution, pausing at each step sufficiently long for the moon's apogee to come up to the line joining the centers of the earth and sun; when the revolution of the sun shall have been completed the revolution of the moon's apogee will, in like manner, have been finished.

If, instead of supposing the sun to advance by successive steps, we admit his uniform progress, it is clearly manifest that in each revolution of the moon, the apogee of her orbit must advance a certain amount in the direction of the sun's motion, and, in the end, a complete revolution of the moon's apogee will be accomplished under the disturbing influence of the sun's attraction.

We have seen that if the sun were stationary his disturbing power would only go to change the figure of the moon's orbit, leaving the direction of the longer axis undisturbed. The revolution of the sun in a circular orbit by slow degrees accomplishes an entire revolution of the apogee, or of the line of apsides, and thus, in case the line of apsides should perform its revolution in a period which shall be an exact multiple of the period of the sun's revolution, then at the end of one such cycle the moon will have passed through all the changes which can arise from the disturbing influence of the sun. These changes will therefore be strictly periodical, and in the end the moon will return to its first position, and will repeat the same identical changes forever.

We will now consider the solar orbit to be elliptical. This involves, by necessity, a perpetual change in the sun's distance, and as his disturbing power varies in intensity inversely with the square of his distance, it is manifest that this variation in the disturbing force will introduce a corresponding variation into the figure of the moon's orbit. If the sun be supposed to advance toward the earth and the moon, in the direction of the line of apsides, its disturbing power would be exerted to draw the moon further from the earth the nearer the sun approached ; in other language, to increase the magnitude of the moon's orbit and the period of her revolution. This action will be varied in case the sun recede from the earth along the same line, and if this approach and recess were made by successive steps, at intervals sufficiently long to allow the moon's orbit to assume a fixed form, then one vibration of the sun advancing and receding through equal space, would work out a series of changes in the form of the moon's orbit identical with those accomplished by each successive vibration, while in all these changes the direction of the line of apsides would remain fixed.

If now we suppose the advance and recess of the sun to be effected by its revolution in an elliptic orbit, then we shall find the changes of figure in the moon's orbit, just noticed, as due to the sun's change of distance, will be combined with an advance and final complete revolution of the line of apsides; and admitting the figure of the sun's orbit to remain unchanged, and the principal axis of its orbit to remain fixed forever in position, a time will come when the sun will have been presented to the moon in every possible position, and all the changes in the figure of the moon's orbit, and the revolution of the line of apsides of the moon's orbit, due to the revolution of the sun in his orbit, will have been accomplished. The moving bodies return to their primitive points of departure, and a new cycle of changes begins, to be repeated, in the same order forever.

Thus far we have supposed the line of apsides of the sun's orbit to remain fixed in position and unchanged in length. It is manifest that a revolution of the line of apsides of the sun's orbit, definite in period and fluctuations in its length also periodical, would introduce additional fluctuations in the moon's motion, and in the length and position of the principal axis of her orbit. But while we rise in complexity, and while the periods requisite for effecting all these changes expand into ages, we still recognize the great fact that *periodicity* remains, and that in the end, at the termination of a vast cycle, the revolving bodies must return again to their points of departure to repeat the same identical changes through endless ages.

In all our reasoning thus far we have supposed that the three bodies under consideration always lie in the same plane; in other language, that the planes of the orbits of the moon and earth coincide. This, however, is not the case of nature. As we have already seen, the moon's orbit is inclined to the plane of the ecliptic under an angle of about 5°—the line of intersection of the two planes being called the line of the moon's nodes, which line must, of course, always pass through the earth's center.

We shall now proceed to take this inclination into consideration, and ascertain whether the sun's disturbing force has any, and if any, what effect on the position of the line of nodes, and on the inclination of the moon's orbit. For this purpose let us suppose the earth to be stationary, and that the line of nodes of the moon's orbit holds a position perpendicular to the line joining the centers of the earth and sun, and that the moon starts from her ascending node to describe that portion of her orbit lying above the plane of the ecliptic. The power of attraction of the sun will manifestly exert itself in such manner as to cause the moon to deviate from its old orbit and to describe a new orbit, which will lie in all its points a little nearer to the plane of the ecliptic. The moon will not, therefore, rise in this superior part of her orbit as high above the plane of the ecliptic as she did before her motion was disturbed by the sun; and in descending to pass through her node she will clearly reach the plane of the ecliptic quicker than she did when undisturbed, and pass through her node at a point nearer to herself than that occupied by the former node; in other language, the old node comes up to meet the advancing moon, and thus takes up a retrograde motion.

Let us now examine the motion of the moon in that portion of her orbit lying beneath the plane of the ecliptic and most remote from the sun. Here the sun's disturbing influence will be diminished somewhat, in consequence of the increased distance at which it operates, but its effect will manifestly be to cause the moon to descend more rapidly and to reach a lower point beneath the ecliptic than when undisturbed, increasing the inclination of the plane of the orbit, and causing the moon to reach her ascending node at a point earlier than when undisturbed, and thus producing a retrocession or retrograde motion of the line of nodes. Thus it appears that in the long run the sun's disturbing influence will tend to change within certain limits the angle of inclination of the moon's orbit; and, indeed, if the earth were fixed in position, would finally destroy this inclination entirely, reducing the plane of the moon's orbit to absolute coincidence with that of the earth; but as the moon is carried by the earth around the sun, and as the moon's orbit in the course of an entire revolution of the earth is thus presented to the sun at opposite points of the orbit under reverse circumstances, there is a compensation accomplished, so far as the angle of inclination is concerned, and also a partial compensation in the retrogression of the line of nodes of the moon's orbit, but not such as to prevent, in the end, a complete revolution of the moon's nodes in a period which we have already seen amounts to eighteen years and two hundred and nineteen days.

We have thus attempted to present a general account of the effect of a disturbing force. These same principles may be extended yet further, and will give a general idea of the effects produced by the planets and their satellites upon each other.

If we return for a moment to the hypothesis that the earth is the only planet revolving about the sun, the magnitude of its orbit, as well as the length and position of the line of apsides, will remain for ever fixed. If, however, we introduce into our system a new planet revolving in an orbit interior to that of the earth, whatever force is exerted upon the earth by the attractive power of this new planet will go to reinforce the power exerted by the sun; and hence the disturbing influence of the planet will tend to diminish the magnitude of the earth's orbit, and to decrease its periodic time. If the disturbing planet revolve in the same direction with the earth, by applying the reasoning hitherto used we shall find that its effect will be to cause the perihelion point of the earth's orbit to advance and retreat during the revolution of the disturbing body, always leaving, however, a slight preponderance of the advancing movement over the retrograde.

In case the disturbing body revolve in an orbit exterior to that of the earth, then its effect will be to expand the earth's orbit and to increase the periodic time, while the influence exerted upon the position of the line of apsides will, in the long run, produce an advance.

The reasoning hitherto employed with reference to the inclination of the moon's orbit to the ecliptic is directly applicable to the effect produced by any planet upon the inclination of the orbit of any other planet, as referred to a fixed plane. Take the earth for example, and let us consider the effect of any planet either interior or exterior upon the inclination of this plane to any fixed plane. So long as the disturbing body is revolving in that part of its orbit lying below the plane of the ecliptic the tendency of the disturbing force will be to draw the earth from its undisturbed path below the plane of a fixed ecliptic, while this effect will be reversed whenever the disturbing planet shall pass through the plane of the ecliptic, and commence the description of that part of its orbit which lies above this plane.

From the above reasoning it is clearly manifest that as not a solitary planet or satellite is moving undisturbed under the attractive power of its primary body, not one of the heavenly bodies describes rigorously an elliptic orbit, nor does the line joining the sun with any planet sweep over precisely equal areas in equal times, neither are the squares of the periodic times of the planets exactly proportioned to the cubes of their mean distances from the sun. In short, every law of Kepler, whereby perfect harmony seemed to be introduced among the heavenly bodies, is now seen to fail in consequence of the law of universal gravitation, and we find ourselves surrounded by a problem of wonderful grandeur, but of almost infinite complexity. Before this problem can be fully solved we must measure the distance which separates every planet from the sun, and which divides every satellite from its primary; we must weigh the sun and all his planets and every satellite; we must determine the exact periods of revolution of each of these revolving worlds; and when all this is accomplished, to trace out the reciprocal influences of each upon the other demands powers of reasoning far transcending the abilities of the most powerful genius, and hence the mind must either forego the resolution of this problem, or prepare for itself some mental machinery which shall give to thought and reason the same mechanical advantages which are obtained for the physical powers of the body by the invention and construction of the mighty engines of modern mechanics.

This has actually been accomplished in the discovery and gradual perfection of a branch of mathematics called the *infinitesimal analysis.* Up to the time of Newton, the mind employed alone the reasoning of geometry in the examination and discussion of the problems presented in the heavens. Even Newton himself was content to publish to the world the results of the application of the law of gravitation to the movement of the planets and their satellites under a geometrical form, exhibiting, in the use of these old methods, a sort of gigantic power which has ever remained as a monument of his wonderful ability.

He was, however, fully conscious of the fact, that the mind demanded for its use, in a full investigation of the physical universe—in the pursuit of these flying worlds, journeying through space amid such a crowd of disturbing influences—a far more subtle, pliable, and powerful mental machinery than that furnished in the cumbrous forms of geometrical reasoning. Conscious of this want, the genius of Newton supplied the deficiency, and gave to the world the *infinitesimal analysis*, which, as improved and extended by the successors of the great English philosopher, has enabled man to accomplish results which seem to place him almost among the gods.

The plan of our work does not permit any attempt to explain the nature and powers of this new method of reasoning. We can only illustrate imperfectly the difference between the use of geometry and analysis. The demonstration of a problem by geometry demands that the mind shall comprehend and hold the first step in the train of reasoning, then, while the first is held, the second must be comprehended, and while intently holding these two steps, the third must be mastered and held, while the mind advances to the fourth step; thus progressing with a constantly accumulating weight oppressing the attention, and tending to crush and destroy further effort to advance, till, finally, the steps become so numerous and complex that only those possessed of a genius of surpassing vigor are able to reach in safety the last step, and thus grasp the full demonstration of the problem. Such is the reasoning of geometry. That of analysis is entirely different. Here the great effort is put forth to master fully and perfectly the conditions of the problem, and then to fasten upon the problem thus mastered the analytical machinery demanded in its resolution. This once accomplished, the mind puts forth its energy and accomplishes the first step, and may there stop and rest, in the full confidence that what has been gained can never be lost. Days, even months may pass, before the problem be resumed, but in this lapse of time there is no loss, and the investigation may be taken up precisely where it was left off; and so one step after another may be taken, each dependent on the other, but each in some sense stereotyped as the mind advances, and remaining fixed without the putting forth of any mental effort to retain it. In short, geometry demands a vigor of mind sufficient to grasp, and hold at the same instant, every link in the longest and most complex chain of reasoning, while analysis only requires a power of genius sufficient to deal with individual links in succession; thus, in the end, reaching the conclusion by short and comparatively easy mental marches.

CHAPTER XI.

INSTRUMENTAL ASTRONOMY.

Method for Obtaining the Mass of the Sun.—For Getting the Mass of a Planet with a Satellite.—For Weighing a Planet having no Satellite.—For Weighing the Satellites.—Planetary Distances to be Measured.—Intervals between Primaries and their Satellites to be Obtained.—Intensity and Direction of the Impulsive Forces to be Determined.—These Problems all Demand Instrumental Measures.—Differential Places.—Absolute Places.—The Transit Instrument.—Adjustments.—Instrumental Errors.—Corrections Due to Various Causes.—American Method of Transits.—Meridian Circle.—The Declinometer.

The general reasoning presented in the preceding chapter can only be reduced to exact application after having obtained the numerical values of the quantities demanded in the investigation. The mathematician may assume these quantities at his pleasure, and with the assumed weight of his sun, and planets and satellites, and with their assumed distances, and with the assumed directions and intensities of the impulsive forces, he may master, by analytical reasoning, all the circumstances attending the revolutions of these supposed worlds, and thus trace their imaginary history for ages, either past or future.

This is the work of the pure mathematician. The physical astronomer takes up the general mathematical reasoning thus perfected, and to employ it in writing out the history of the real bodies constituting the solar system, he must measure the actual distances between the sun and the planets, and the distances from each primary to its satellites; he must weigh exactly the sun and each of the planets and satellites, and he must measure in some way the direction and intensity of the impulsive forces by which the planets and their satellites were projected in their respective orbits.

We shall now proceed to show that the determination of all these quantities depends on exact astronomical measurements, which measurements demand the invention and construction of instruments of the highest order of power, delicacy and perfection.

To weigh the sun and planets.—Let it be borne in mind that the law of gravitation asserts that bodies attract with a force or power directly in proportion to their mass or weight. Hence a sun, weighing twice as much as the central orb of the solar system, would (at the same distance) attract with a double force. The same is true of the earth; and if it were possible to hollow out the interior of the earth until its weight were reduced to one-half of what it is now, its power of attraction would be diminished in the same exact proportion.

Thus, to know with what power the sun or any planet or any satellite attracts a body at a given distance, we are compelled to ascertain the exact weight of the sun, planet, or satellite.

We shall show hereafter that it is possible to reach an approximate value of the weight of the earth in pounds avoirdupois, but for our present purpose it will be sufficient to state that the weight of the earth is well represented by the intensity of its power of attraction at a unit's distance from its center. For this unit of length we will take the *earth's radius*, and hence a body on the surface is attracted by a force or power such as will measure the mass or weight of the earth; but the intensity of any force is measured by the quantity of motion it is capable of generating in a given time. Hence the intensity of the earth's attractive power will be correctly measured by the velocity it impresses on a body free to fall, in, say, one second of time. This is a matter of the simplest experiment, by which it is found that the earth's attractive power generates in one second a motion in a falling body such as to carry it over a space equal to about sixteen feet in one second. In case the earth were twice as heavy, the space passed over by a falling body in one second would be doubled, and so forward in like proportion for any increase of weight.

Having thus found a measure of the weight of the earth in the space passed over in a second of time by a falling body, in case it were possible to transport this body to the surface of the planet Venus (assuming the diameter of Venus and the earth to be equal), then permitting it to fall, and measuring the space over which it passes in one second, this space would hold the same proportion to sixteen feet as the weight of Venus does to that of the earth. If the diameters of the planets are unequal, then we must take into account the fact that the falling body is not at equal distances from the centers of the planets, and that the force of attraction is thus diminished inversely as the square of the distance from the center is increased. Let us attempt to weigh two planets whose diameters are in the proportion of one to two. At the surface of the smaller planet suppose the body falls sixteen feet in one second, while at the surface of the larger planet it passes over sixty-four feet in the same time. In case the diameters were equal this result would show that one body was four times as heavy as the other; but the falling body is twice as far from the center of the large planet as it is from the center of the small one, and hence the force or power of attraction of the large planet is only one-fourth part what it would have been in case the falling body had been brought to within one unit of its center. If, then, with an energy reduced at a distance of two units to one quarter, it causes a fall of sixty-four feet in one second, the entire energy would, at a unit's distance, cause a fall through four times sixty-four feet, or through 256 feet, and hence the weights of the planets under examination are in the proportion of 16 to 256 or 1 to 16.

The train of reasoning here presented may be extended to embrace any given case, and if it were possible to make the experiment of the falling body, as above described, at the surfaces of the sun, planets and satellites, (admitting that we know the diameters of all these bodies,) then would it be possible to ascertain their masses, as compared with that of the earth, taken as a unit.

But it is impossible to pass to the sun and planets for such experimentation, and hence we must devise some substitute which may fall within the limits of practicability. To obtain the relative weights of the sun and earth we have only to call to mind the fact that the moon, under the power of the earth's attraction, is ever falling away from the rectilineal path in which it would fly but for this very power of attraction, while in like manner the earth is ever falling away from the right line in which it would move but for the attractive energy of the sun. Here, then, are two bodies, the earth falling to the sun, the moon falling to the earth;

and in case we could measure the precise distance which each of these bodies falls, under the respective powers of attraction exerted on them, taking into account the effect produced on the two forces by the inequality of the distances at which they operate, we should reach the exact relative weights of the sun and earth. Thus, admitting that the distance at which the sun operates on the earth is 400 times greater than the distance at which the earth operates on the moon, in case the effects were equal the sun would be 160,000 times heavier than the earth, since its power of attraction is reduced by the distance in this exact ratio. But again, admitting that we find, even with this high reduction, the sun's power on the earth is still two and a half times greater than the earth's power on the moon, (as is shown in their respective deflections from a right line in one second of time,) then will the sun be $2\frac{1}{2} \times 160,000$ times heavier than the earth, and this, indeed, as we shall find hereafter, is about the relative weights of these two globes.

To resolve this great problem, then, of weighing the sun against the earth, we must first measure the sun's distance and the moon's distance, and the exact amounts by which the earth and moon are caused to fall away from a rectilineal orbit in one second of time, which measurements demand *instruments* of the highest order.

TO WEIGH AGAINST THE EARTH, A PLANET ATTENDED BY A SATELLITE.—In case any planet be attended by a satellite, if we can measure the precise distance separating these two bodies, and determine the period of revolution of the satellite, we can thence derive the weight of the planet as compared with that of the earth. To fix our ideas, let us suppose Jupiter's nearest satellite to be as far from Jupiter as our moon is from the earth, and to perform its orbital revolution in the same exact time occupied by the moon. This would prove Jupiter to be just as heavy as the earth. But suppose now that at equal distances Jupiter's moon revolves ten times as rapidly as the earth's moon, this fact proves that Jupiter must be one hundered times as heavy as the earth. This is evident from what we have already said, that the centrifugal force in any revolving body increases as the *square* of the velocity; and as the moon of Jupiter is now supposed to revolve ten times as fast as our moon its centrifugal force will be one hundred times as great as that of the earth's moon; and hence Jupiter's attraction to counterbalance this tendency to fly from the center must be one hundred fold greater than that of the earth. This is on the hypothesis of equal distances. But if Jupiter's moon be supposed to be twice as remote from its primary, and to revolve ten times as rapid as our moon, then will it be demonstrated that Jupiter is one hundred times heavier, on account of the square of the velocity of the revolving moon, but this weight must be multiplied by the square of two on account of the double distance at which it acts. Hence, under these circumstances Jupiter would be 400 times as heavy as the earth.

Thus, to determine the weight of a planet in terms of the earth's weight as unity, we must learn the exact distance and periodic time of our moon, and also the interval by which the planet and its moon are separated, as well as the period of revolution of the satellite, all of which again demand the use of *instruments* of a high order of accuracy and delicacy.

TO WEIGH A PLANET HAVING NO SATELLITE.—Three of the planets, viz.: Mercury, Venus, and Mars, so far as known, are not accompanied by a moon. The preceding method of obtaining the mass or weight will not apply to either of these planets. It is only after acquiring a very exact knowledge of the movements of the planets whose masses may be derived from their satellites that it becomes possible to determine the weights of the remaining planets. Let us suppose that the earth alone revolved around the sun, and that its orbit was perfectly determined. In an exterior orbit of known dimensions let us place the planet Mars. This will at once modify the former orbit of the earth, and the change will depend, in quantity, upon the mass of the new planet; and in case it became possible to measure these changes, their values will give the weight of the body producing them.

The same hypothesis remaining with reference to the earth's orbit, we may imagine the new planet to revolve in the orbit of Venus, interior to that of the earth, and the same kind of investigation will lead to the determination of the mass of this interior planet.

We shall see hereafter that certain periodical comets, favorably located, furnish the means of corroborating the results reached by the above train of reasoning, by the data their perturbations furnish for reaching the mass of the planet producing these effects.

TO WEIGH THE SATELLITES.—The effect produced by the moon on the earth in causing the figure of its orbit to sway to and fro under the moon's attractive power furnishes again the data whereby the moon's mass may be determined. In the case of many satellites to the same planet, their effects on each other being carefully determined, furnish the means of computing their masses. This, however, is a difficult problem, and one in which a solution has been effected only in the system of Jupiter. The masses of the satellites of the other superior planets have as yet not been obtained with any reliable certainty.

We have thus presented methods by which the masses of the sun, planets and satellites may be obtained, provided certain measurements can be made, which measurements demand the aid of powerful and accurate *instruments.*

The *distances* separating the sun and planets, and separating the primaries and their satellites, must be obtained before we can trace the history of any one of these revolving worlds. We have already explained the processes by which the earth's distance from the sun may be obtained by the use of the phenomena attending the transit of Venus. This problem again demanded *instrumental measurement.* Admitting the earth's distance from the sun to be known, Kepler's third law will give the distances of all the planets of our system, provided we have obtained their periods of revolution around the sun. The method of obtaining the periodic times has also been explained, and in this process *instrumental measurements* are demanded.

In like manner, to reach the periods and distances of the satellites, their elongations, occultations and eclipses, must be carefully measured and noted, demanding *instruments* of a high order.

To trace a planet or satellite, in addition to the quantities already pointed out, we have seen that we must know the *intensity of the impulsive force* by which it was projected in its orbit. But we have seen that the intensity of any impulse is measured by the velocity it is capable of producing in a unit of time. Admitting, then, that we know the distance of a planet from the sun, and its period of revolution, we know the velocity with which it moves, in case its orbit be circular. The earth, for example, in 365¼ days accomplishes a journey round the sun in a circle whose diameter is 190,000,000, and whose circumference is equal to this quantity taken 3.14156 times. Hence, by dividing the number of miles traveled in the entire circuit by the number of days occupied in the journey, we have the rate per diem, or velocity. Dividing the space passed over in one day by 24, we have the rate per hour, and finally may obtain the rate per second. If the orbit be not circular, we can always find a circle which, for a very short distance, will coincide with an elliptic or other curve, and on this circle we may suppose the planet to move for a very short time, as one second, with uniform velocity, and the space passed over in this unit will again measure the intensity of the impulsive force at this part of the orbit. Here again we have presupposed a knowledge of the magnitude and figure of the elliptic orbit before the intensity of the impulsive force can be reached, and to determine these quantities instrumental measurement is demanded, requiring *instruments* of great perfection.

The last quantity demanded by the mathematician in writing out the history of a planet moving in space is the *direction* of the impulsive force projecting it in its orbit. This is readily obtained when we shall have learned the exact direction of a line tangent to any point of the planetary orbit; for the direction of the impulse must always be tangent to the curve described by the body set in motion. If we join the planet with the sun by a right line, this line will form an angle with tangent to the planetary orbit; and we shall find hereafter that the nature of the orbit will depend upon the value of this angle, or in other language, on the direction in which the impulse is applied.

Thus we find that not a single quantity of the five required to determine the circumstances of motion of a body revolving under the laws of motion and gravitation can be reached without *instrumental measurement;* so that our entire knowledge of the physical universe hangs at last on the accuracy and perfection of the *instruments* which have been invented and constructed for making these measures, a fact which elevates *instrumental astronomy* to a position of the highest dignity and importance.

The measures demanded in instrumental astronomy are divided into two great classes. In the *first class* all the measures of position are *absolute,* that is, a star or planet whose place is thus determined is located on the celestial sphere, and fixed for the moment in position by a measure of its distance, say, from the north pole of the heavens along the arc of a great circle, and also its distance measured on the equinoctial from the vernal equinox, or from some other fixed points which may have been selected. In the *second class* all the measures are *relative* or *differential;* that is, an interval between two points in close proximity is determined. To this class belong the measures of the diameters of the sun and moon and planets; the elongations of the satellites from their primaries; the measures of the transits of Venus and Mercury across the disk of the sun; the measures of the solar and lunar spots; the distances between the double and multiple stars; in short, all those measures involving mere differences of position.

Each of these classes of measures demands its own peculiar and appropriate instruments, each of them involving the data required in the solution of the sublimest problems of celestial science.

We shall now proceed to exhibit an outline of the structure of a few of the most important instruments belonging to these two classes, only for the purpose of presenting the extraordinary difficulties which must be met and conquered in the seemingly simple mechanical problem of fixing the place of a star in the celestial sphere.

For the purpose of giving position to the heavenly bodies astronomers refer them to the surface of a celestial sphere, whose poles are the points in which the earth's axis prolonged pierces the sphere of the fixed stars. To determine a point on the surface of any sphere we must fix its distance on the arc of a great circle from the north pole, and we must also know the distance of the meridian line on which it is located, from a fixed meridian.

Astronomers have chosen for their prime meridian that one which passes through the vernal equinox, and as the celestial sphere revolves to our senses with uniform velocity once in twenty-four hours, the vernal equinox will come at the end of this period to the meridian of the place from whence it started. Any object, therefore, which crosses the meridian of a given place an hour later than the vernal equinox has its place fixed somewhere on the circumference of a known meridian, or hour circle. If at the same time its distance from the north pole can be determined, its position on the celestial sphere will be positively defined by these two elements. As we have already seen, the vernal equinox is the point in which the great circle of the heavens, cut out by the indefinite extension of the plane of the earth's orbit, intersects the equinoctial circle, or that circle cut from the celestial sphere by the indefinite extension of the plane of the earth's equator. If the vernal equinox were absolutely fixed, and if in that point a star were located, this star would revolve with all the other stars of the heavens once in twenty-four sidereal hours. To mark the movement of this vernal equinox astronomers employ the sidereal clock, whose dial is divided into twenty-four hours, and which, when perfectly adjusted, will mark 0h. 0m. 0s. at the moment the vernal equinox is on the meridian of the place where the clock is located. All points on the celestial sphere will pass the meridian necessarily at intervals of time marking the position of the hour circle in which they are located, relative to the prime meridian passing through the vernal equinox. These intervals of time which elapse between the passage of the vernal equinox across the meridian of a given place, and the passage of any heavenly body across the same meridian, are called *right ascensions*. Thus a star which follows the vernal equinox, after an interval of 2h. 10m. 20s., as marked by a perfect sidereal clock, has a right ascension of 2h. 10m. 20s.

Thus, to fix the place of any heavenly body on the celestial sphere, two instruments have been devised, the one having for its object to measure *north polar distances*, while the other is employed in the measurement of *right ascensions;* the first of these is denominated a *mural circle*, while the second is called a *transit instrument.*

We shall first consider the principles involved in the construction of the transit instrument. This instrument consists of a telescope mounted upon an axis perpendicular to the axis of the tube of the telescope. This perpendicular axis terminates at each extremity in two pivots of equal size and perfectly cylindrical in form. To give support to this instrument a solid pier of masonry is built, resting upon a firm foundation, and isolated from the surrounding building. On the upper surface of this stone pier two stone columns are placed, whose centers are separated by a distance equal to the length of the axis of the transit; on the tops of these columns metallic plates are fastened, to which metal pieces are attached, cut into the shape of the letter Y. If in these Y's the pivots of the transit be laid, in case the axis be precisely level, and lying due east and west, then the axis of the telescope, or visual ray, being carried around the heavens, by revolving the instrument in its Y's, will describe a meridian line which will pass through the north pole of the heavens. If this meridian line could be rendered visible it would be possible to note the passage of any star or other heavenly body across this visible meridian. This cannot be accomplished directly, but the same end is reached by stretching a delicate filament of spider's web across the center of a metallic ring, and placing it in the focus of the eye-piece of the telescope; when this spider's web is lighted up by a lamp, through a suitable orifice, it is seen as a delicate golden line of light stretching across the field of view, and resting on the dark background of the heavens. Revolving the ring which bears the spider's web, we may bring this web to coincide, throughout its entire length, with a true meridian line, and thus, in reality, we procure for ourselves a visible meridian quite as perfect for our purposes as though it were an actual line of light, sweeping from north to south across the celestial sphere. To render visible the axis of the telescope, or to direct the visual ray, another spider's web is stretched across the field of view, in direction perpendicular to the first, and precisely in the center of the field, so that by their intersection these spiders' webs form a point of almost mathematical minuteness.

Let us now examine what is demanded in the construction of the transit to render it an instrument perfect in performance. The *object-glass* and the *eye-piece*, forming the optical portion of the telescope, should be perfect in their figure and adjustments; the tube in which they are placed should be perfectly rigid and inflexible; the optical axis or *line of collimation* should be exactly perpendicular to the horizontal axis on which the instrument revolves; the *pivots* should be exactly equal to each other, and precisely cylindrical in form; the *horizontal axis* should lay in a direction exactly east and west, and should be absolutely level. In connection with the transit we require a perfect time-keeper. The clock, when properly adjusted, will mark 0h. 00m. 00s. at the moment the vernal equinox is seen to pass the visible spider's line meridian of the telescope; it must then move uniformly during the entire revolution of the heavens, and mark the zero of time again when the vernal equinox returns to the meridian line. Such are the mechanical demands required in the construction and use of the transit and clock; but to obtain a perfect result the observer is required to perform his part in the operation; he must note the *exact instant* at which the vernal equinox passes the visible meridian, so as to set his sidereal clock; this being accomplished, to obtain the right ascension of any heavenly body he must seize the *precise mo-*

ment, as marked by his clock, at which the center of the object under observation passes the meridian.

To obtain, then, the element of right ascension, required to fix the place of a heavenly body at a given moment, we require a perfect transit, perfectly adjusted, a perfect clock, perfectly rated, and a perfect observer, with a perfect method of subdividing time into minute fractions. Not one single one of these demands can ever be met. Even admitting the possible construction of a perfect instrument, every change of temperature will effect certain changes in the material of which it is composed. No two pivots can possibly be made exactly equal, nor precisely cylindrical in form; and should the observer succeed in placing the axis of his transit so as to lie east and west, as well as horizontal, it will not remain in this position for even a single hour of time. If the clock be adjusted so as to mark the exact zero of time, and to move off with a uniform rate, this rate will soon sensibly change, and must be carefully watched even from hour to hour. The observer himself is but an imperfect and variable machine, utterly incapable of marking the exact moments required, his work being subject to errors, whose values fluctuate from day to day; and to add to all these difficulties, the atmosphere which surrounds the earth not only possesses the power of diverting the rays of light from their rectilineal path, but because of its constant fluctuations and changes produces a tremulous or dancing motion in the stars under observation which, to a certain extent, renders it impossible to do exact work, even with perfect instruments and perfect observers, could such be found.

We will now examine the instrumental means required to determine the second element, *the north polar dis-*

tance, which is demanded in fixing the place of a heavenly body. For this purpose let us suppose a metallic circle to be permanently fastened to the horizontal axis of the transit, having its center in the central line of the axis, and its plane perpendicular to this line. Let us suppose the rim of this circle to be divided into degrees, minutes and seconds of arc, and this division to have been perfectly accomplished. Let us direct the transit telescope precisely to the north pole of the heavens, and when thus directed let us fix upon the stone pier a permanent mark or pointer, directed to the zero point on the divided circle. As we turn the transit away from the north pole toward the south the zero point on the circle will in like manner leave the fixed pointer, and thus the distance from the north pole to any object to which the telescope may be directed will be read on the divided circle from the zero round to the division to which the pointer directs. Such an instrument is called a *meridian circle*. Here again new mechanical difficulties present themselves. The centering of the circle, the perfection of the divisions upon its circumference, are matters which cannot be accomplished with absolute accuracy; and even if this were possible, they are liable to changes to which all material is subjected at every moment. The same is true of the stability of the pointer, any change in whose position must involve an error in the measured north polar distance.

Thus far we have supposed that in our celestial sphere we have two *fixed* points of reference, namely, the *vernal equinox* and the *north polar point*. Unfortunately for the observer, and to increase the difficulties by which he is surrounded, neither of these points remains absolutely fixed, and even their rate of movement is not uniform; and thus one difficulty rises above another, culminating in the fact that even the light whereby objects become visible does not dart through space with infinite velocity, but wings its flight with a measurable speed, which, when conjoined with the speed of the earth's revolution in its orbit, sensibly changes the apparent place of every object under examination. Add to this long catalogue of difficulties the fact that the earth's rotation on its axis is rapidly revolving the observer, his instruments and observatory, at a possible rate of a thousand miles an hour, and some idea may be formed of the embarrassments under which astronomers are compelled to work out the resolution of the great problems of the heavens.

Having presented this array of difficulties, we shall not undertake to show in every instance by what precise means they are overcome. So far as regards the movement of the vernal equinox and the north pole, the most extended and elaborate observations have been made through a long series of years by the best instruments and the most skillful observers, and these have been reduced and discussed by the most able mathematicians, until by different methods astronomers have reached to so perfect a knowledge of the values of the errors due to these two causes that it seems as though no greater approximation to accuracy can be made by the same methods. That correction due to the movement of the vernal equinox is called *precession;* that due to the movement of the north pole is called *nutation*, of which we shall give a more accurate account hereafter. The error arising from the velocity of light, combined with the orbital and rotary motion of the earth, is called *aberration*. This subject will also be treated hereafter. Of the three principal instrumental errors, that arising from want of exact perpendicularity in the position of the axis of the telescope to the horizontal axis of the transit is called the *collimation* error, and may be detected and measured by mechanical means; its effect is to cause the visual ray to pierce the heavens east or west of the true meridian, and in the revolution of the transit this point of piercing will describe a small circle of the sphere, instead of the great meridian circle, which it ought to describe. The error arising from a failure to place the transit axis in a truly horizontal position is called the *level* error; its effect is to cause the visual ray to pierce the heavens east or west of the true meridian, and the point of piercing, by the revolution of the transit on its axis, will describe a great circle of a sphere inclined to the true meridian, under an angle equal to that which the axis of the transit makes with the horizon, or equal to the level error. The failure to place the axis of the transit precisely east and west gives rise to what is called the *azimuthal* error. This causes the visual ray or line of collimation of the telescope to pierce the heavens on the true meridian only when directed to the zenith. This point of piercing, by revolving the transit on its axis, describes a great circle, which departs from the true meridian at the zenith, under an angle precisely equal to the azimuthal error. Methods have been devised for measuring these various errors, and for computing their effect upon the apparent places of the heavenly bodies.

The rays of light by which every object is rendered visible, as we have already stated, on entering the earth's atmosphere are bent from their rectilineal path, giving rise to a source of error called *refraction*. The laws governing the direction of the light, as affected by the atmosphere, have been carefully studied, so that at present it is possible to compute with great exactitude the change of place of any object under observation due to the effects of refraction. The flexure of the tube of the telescope, under the various circumstances by which it may be surrounded, have been thoroughly investigated, while the exact figure of the pivots of the axis has been subjected to the most rigorous mechanical tests; in short, all the mechanical deficiencies in the instrument have occupied the attention of many of the best minds for the past two hundred years, and thus slow but steady advances in accuracy have been accomplished.

To remedy the errors arising from the perturbations of the atmosphere, as well as those arising from personal error in the observer, in seizing the moment of transit across the visible meridian line, several spider's lines, commonly called *wires*, parallel to each other, have been introduced into the focus of the eye-piece of the transit, and thus the instant at which the star passes each one of these wires being noted as accurately as possible, the average of all gives a better result than could have been obtained from any one wire.

To remedy the defects arising from the imperfect divisions, from imperfect centering, and from changes of figure in the circle from whence the north polar distances are read, it is usual to have four pointers, and even sometimes six, by means of which the north polar distance is read in as many places on the divided circle, the average of all giving a better result than any one reading. These pointers, as we have named them, are in reality powerful microscopes, permanently fixed in the heavy stone pier, on which the instrument rests, and having their visual ray fixed by the intersection of spiders' webs, as in the principal telescope.

From these *instrumental* imperfections we pass to those which belong to the clock, and here again we are compelled to work with an imperfect machine. No clock has ever been made which can keep *perfect* time, and the great object of the observer is to learn the peculiarities of his clock, to determine its deviations from absolute accuracy, and to be able to mark these deviations, if possible, from minute to minute.

The observer, having mastered all the sources of error above described, next comes to the consideration of his own *personal* deviations from accuracy in attempting to mark the moment at which a star crosses his visible meridian. To observe the transit of a star across the meridian, he places himself at the transit instrument, enters in his note-book the hour and minute from the face of the clock, then fixing his eye through the telescope upon the star, and counting the beats of the pendulum, he follows the star as it slowly advances to the meridian wire. Between some two beats thus counted the star crosses the wire. The observer holds in his mind, as well as he can, the star's position at the close of the beat before the passage, and at the close of the next beat after the passage, and mentally subdividing this space passed over in one second into ten equal parts, he estimates how many of these parts precede the passage of the star across the meridian, and these parts are the fractions or tenths of a second, which mark the time of transit. Thus he adds to the entry in his note-book already made the number of beats of the pendulum and also the fractions of a second above obtained, and thus the time of transit is obtained, *approximately* to the tenth part of one second of time.

In the method of observing transits just explained so

many things are demanded of the observer that his attention cannot be given exclusively to the determination of the moment of transit; he must keep up the *count* of the clock beat; he must hold in his mind the *interval* passed over by the star from one beat to the next during the transit; he must divide this space by estimation into *tenths;* he must assign the *number of tenths* which precede the transit; he must enter the seconds and tenths in his note-book, keeping up the count of the beats of the pendulum, and thus pass from one wire to the next successively through all the system of wires, so that in this multiform effort his powers of attention are taxed beyond what they are able to bear, and it is only by long practice that any valuable results are ever reached. The observer also finds that his modes of observation often lead him into false habits. He may mark the time from his own *mental count* of the beat, rather than from the *sound* of the beat itself, or he may find himself running into the habit of fixing his tenths of seconds predominantly in one or two portions of the scale of tenths. It is manifest that in a thousand observations the tenth of a second on which the transit falls ought to be uniformly divided among the whole number. Thus there should be a hundred observations in which the time of transit should fall on the first tenth of a second, a hundred observations in which the time should fall on the second tenth, and so on for each of the tenths. But an observer may find when he comes to examine a thousand of his observations that two or three hundred are entered as falling on the third tenth, and three or four hundred as falling on the seventh tenth. This only demonstrates that he has fallen into habitual error, due to the fact that he is compelled to *estimate.* In attempting to escape from this particular error, and finding himself too much attached to one portion of his scale of tenths, he is very likely to fall into the other extreme, and thus he finds himself a variable instrument, always imperfect, even in these legitimate sources of error.

By studying his own peculiarities more rigorously, and comparing himself with others, it will be found that in case the two persons compared could at the *same* time look through the *same* telescope at the *same* star coming up to cross the *same* meridian wire, each attempting to note the moment of passage, by listening to the beat of the *same* clock, the recorded times would differ, one of the observers being uniformly in advance of the other. Should this experiment be repeated, at the end of a month, with every possible precaution, the difference between the two observers will in general be found to change, demonstrating that one or the other or both have varied in this particular, and that an inter-comparison of their observations now made by the former difference would produce inaccurate results. This difference is what is denominated technically *personal equation*, and is supposed to arise from the fact that time is really an element in the operation of the senses: that two persons listening to the same sound, as the sharp crack of a pistol, the sense of hearing of the one may perform its office of conveying this sound to the brain more rapidly than the other, and that the same may be asserted of the sense of sight.

For the purpose of comparing the observations of different astronomers, it becomes necessary to determine the peculiarities of each, and it would be a matter of great importance if it were possible to fix some absolute standard to which all observations might be reduced. This is accomplished, so far as the three *instrumental* errors and the *clock* error are concerned, by actually applying a correction which reduces each observation to what it would have been in case none of these errors had existed. The same may be said of the correction applied for refraction and for aberration. As to precession, the position of the equinoctial point, supposing it to move with its mean or average velocity, is always given for the epoch to which the observation is referred. The observations are also reduced for *parallax* whenever this element becomes sensible, and are thus recorded as though the observer were located at the center of the earth. To accomplish the inter-comparison of observations made at different observatories, there yet remains the reduction due to *difference of longitude*, and that depending upon the *personal peculiarities* of the observers.

The high demand for accuracy in instrumental observation can only be fully appreciated by those actually engaged in the computation of the places of the heavenly bodies. Observations are valuable in the ratio of the *squares* of their probable errors: that is, if one set of observations can be produced in which the probable errors remain among the tenths of seconds of time, while in another set of observations the errors are driven into the hundredths of seconds, or are but one tenth part as large as the former, then the second set will be a *hundred* fold more valuable than the first. This principle applies to all observations, but there are some distances so great and some motions so slow that even the best and most delicate methods of observation hitherto applied fail altogether to measure the one or to appreciate the other. This remark is especially true when applied to the distance and movements which are found in the region of the fixed stars. Among these remote objects, while in some instances the motion is sufficiently rapid to be detected and approximately measured, even in a single year, in other instances, and by far the larger number, these motions are so slow that they must accumulate for hundreds of years to become appreciable and measurable by the most refined and perfect instruments hitherto prepared by human skill.

In the three great departments of astronomy there is but one in which there is much hope for increased facility and accuracy. The great laws of motion and gravitation are no doubt perfectly determined. The mathematical formulæ whereby these laws are applied to the circumstances of motion of the planets and their satellites are now brought to great simplicity and perfection; and if it were possible to give to the physical astronomer *perfect* data, he would be able to obtain *perfect* results. We know by geometry that the area of a rectangle is the product of its base by its altitude. This rule or formula is absolutely accurate, and whenever we wish to apply it to determine the area of any particular rectangle we must first accomplish the mechanical measurement of the length of the base and altitude. To do this perfectly is impossible, but approximate results may be reached of greater or less precision in proportion to the accuracy of the instruments employed, and the time and pains expended upon the work. Thus one measure may reduce the probable errors to one hundredth of an inch, while in another the error may only reach one thousandth of the same unit.

In like manner the theory and formulæ of physical astronomy are nearly, if not quite perfect, while, however, the observations whence we derive the data to be used in computation are, as we have seen, comparatively imperfect. The author of this work has attempted to contribute something to the accuracy and facility of astronomical observation.

The following is a brief account of the circumstances attending the invention of this new mode of observation, now known as

THE AMERICAN METHOD OF TRANSITS.—In the autumn of the year 1848, the late Professor S. C. Walker, then of the United States Coast Survey, was engaged with me at the Cincinnati Observatory in a series of observations, having for their object the determination of the difference of longitude between the observatories of Philadelphia and Cincinnati. In comparing our clocks or chronometers with those of Philadelphia, an observer at Philadelphia listening to the clock-beat touched the magnetic key of the telegraph wire at every beat, and we received at Cincinnati an audible *tick* every second of time, which was carefully noted, and thus our clocks were compared. There were two sources of error in this method of comparison, arising from an imperfect imitation of the clock-beat by the Philadelphia operator, also from our noting the arrival of that beat in Cincinnati. On the 26th of October, 1848, Professor Walker, while conversing on this subject, first presented to me the mechanical problem of causing the clock to send its own beats by telegraph from one station to the other, or what amounted to the same thing, the problem of *converting time into space*, as already explained; for in case the clock could send its own beats by telegraph, and these beats could be received on a uniformly flowing time scale, the star transit could be also sent by telegraph, and received on the same scale; and thus a new method of transits would at once spring from the resolution of the first mechanical problem. I was informed by Professor Walker that the problem had already been presented to others, but, so far as he knew, had never been solved. The full value of the idea was at once appreciated; and on the same evening a common brass clock, the only one then in the observatory, was made to record its own beats by the use of the electro-magnet on a Morse fillet.

The problem once solved, nothing more remained than to elaborate such machinery as would render it possible to apply this new discovery or invention to the delicate and positive demands of astronomical observations.

It is well known that signals are transmitted along a line of telegraphic wire by closing or by breaking the wire circuit over which the electricity passes from pole to pole of the battery. The finger of the telegraphic operator, by touching a magnetic key, "breaks or makes" the circuit, and thus either interrupts or starts the flow of electricity. The problem of causing a clock to record its beats telegraphically was then nothing more than to contrive some method whereby the clock might be made (by the use of some portion of its own machinery) to take the place of the finger of the living, intelligent operator, and "make" or "break" the electric circuit. The grand difficulty did not lie in causing the clock to play the part of an automaton in this precise particular, but it did lie in causing the clock to act automatically, and at the same time perform perfectly its great function of a time-keeper. This became a matter of great difficulty and delicacy; for to tax any portion of the clock machinery with a duty beyond the ordinary and contemplated demands of the maker, seemed at once to involve the machine in imperfect and irregular action. After due reflection it was decided to apply to the *pendulum* for a minute amount of power, whereby the making or breaking the electric circuit might be accomplished with the greatest chance of escaping any injurious effect on the going of the clock. The principle which guided in this selection was, that we ought to go to the prime mover (which in this case was the clock weights, and which could not be employed,) and failing to reach the prime mover, we should select the nearest piece of mechanism to it, which in the clock is the pendulum. A second point early determined by experiment and reflection was this: that the making or breaking of the circuit must be accomplished by the use of mercury, and not by a solid metallic connection. The method evolved and based on these two principles is the one which has been in use now for more than ten years in the Cincinnati Observatory.

The simplest possible method of causing the pendulum to "make" the circuit may be described as follows:

Attach to the under surface of the clock pendulum with gum shellac a small bit of wire bent thus, ⌐¬ then right and left of the point over which the pendulum vibrates when lowest place two small globules of mercury, into each of which there shall dip a wire from the poles of the battery. Now, as the pendulum swings over the globules of mercury, the two points of the attached wire will finally come, for one moment, to dip in the mercury cups, and thus make a momentary *bridge*, over which the current of electricity may pass from pole to pole. This method, among others, having been tried, was soon abandoned as uncertain and irregular in its results; and the following plan was adopted:

A small cross of delicate wire was mounted on a short axis of the same material, passing through the point of union of the four arms constituting the cross. This axis was then placed horizontal on a metallic support, in Y's, where it might vibrate, provided the top stem of the cross could be in some way attached to the pendulum of the clock, and the "cross" should thus rise and fall at its outer stem as the pendulum swings backward and forward. The metallic frame bearing the "cross" also bore a small glass tube bent at right angles. This was filled with mercury, and into one extremity one wire from the pole of the battery was made to dip; the other wire was made fast by a binding screw to the metallic stand bearing the "cross," and thus every time the "cross" dipped into the mercury in the bent tube, the electricity passed through the metallic frame, up the vertical standards bearing the axis of the cross, along the axis to the stem, and down the stem into the mercury, and finally through the mercury to the other pole of the battery. Thus at every swing of the pendulum the circuit was made, and a suitable apparatus might, by the electro-magnet, record each alternate second of time.

The amount of power required of the pendulum to give motion to the delicate wire-cross was almost insensible, as the stems nearly counterpoised each other in every position. Here, however, there was great difficulty in procuring a fibre sufficiently minute and elastic to constitute the physical union between the top stem of the cross and the clock pendulum. Various materials were tried, among others a delicate human hair, the very finest that could be obtained, but this was too coarse and stiff. Its want of pliancy and elasticity gave to the minute "wire-cross" an irregular motion, and caused it to rebound from the globule of mercury into which it should have plunged. After many fruitless efforts, an appeal was made to an artisan of wonderful dexterity; the assistance of the *spider* was invoked; his web, perfectly elastic and perfectly pliable, was furnished, and this material connection between the wire-cross and the clock pendulum proved to be exactly the thing required. In proof of this remark I need only state the fact that one single spider's web has fulfilled the delicate duty of moving the wire-cross, lifting it, and again permitting it to dip into the mercury every second of time for a period of more than three years! How much longer it might have faithfully performed the same service I know not, as it then became necessary to break this admirable bond, to make some changes in the clock. Here it will be seen the same web was expanded and contracted each second during this whole period, and yet never, so far as could be observed, lost any portion of its elasticity. The clock was thus made to close the electric circuit in the most perfect manner; and inasmuch as the resistance opposed to the pendulum by the "wire-cross" was a constant quantity and very minute, thus acting precisely as does the resistance of the atmosphere, the clock, once regulated with the "cross" as a portion of its machinery, moved with its wonted steadiness and uniformity. Thus one grand point was gained. The clock was now ready to record its own beats automatically and with absolute certainty, without in any way affecting the regularity of its movement. It was early objected to the mercurial connection just described, that in a short time the surface of the mercury would become oxidized, and thus refuse to transmit the current of electricity; but experiment demonstrated that the explosion produced by the electric discharge at every dip into the mercury threw off the oxide formed, and left the polished surface of the globule of mercury in a perfect state to receive the next passage of the electricity.

So far as known, all other methods are now abandoned, and the mercurial connection is the only one in use.

THE TIME SCALE.—The clock being now prepared to record its beats, accurately and uniformly, the next important step was to obtain, if possible, a uniformly moving time-scale, which should be applicable to the practical demands of the astronomer.

In case the fillet of paper used in the Morse telegraph could have been made to flow at a uniform rate upon its surface, the clock could now record its beats, appearing as dots separated from each other by equal intervals. But it was soon seen that the paper could not be made to flow uniformly; and even had this been possible, a single night's work would demand for its record such a vast amount of paper that this method was inapplicable to practice. After careful deliberation, the "revolving disk" was selected as the best possible surface on which the record of time and observation could be made. The preference was given to the disk over the cylinder for the following reasons:—The uniform revolution of the disk could be more readily reached. The record on the disk was always under the eye in every part of it at the same time, while, on the revolving cylinder, a portion of the work was always invisible. One disk could be substituted for another with greater ease, and in a shorter time; and the measure of the fractions of seconds could be more rapidly and accurately performed on the disk than on the cylinder.

After much thought and experiment it was decided to adopt "a make circuit" and "a dotted scale" rather than a "break circuit" and a "linear scale;" and I think it will be seen hereafter that in this selection the choice has been fully justified in practice. These points being settled, the mechanical problems now presented for solution were the following: First, To invent some machinery which could give to a disk of, say, twenty inches diameter, mounted on a vertical axis, a motion such that it should revolve uniformly once in each minute of time; and, second, To connect with this disk the machinery which should enable the clock to record on the disk each alternate second of time, in the shape of a delicate round dot. Third, The apparatus which should enable the observer to record on the same disk the exact moment of the transit of a star across the meridian, or the occurrence of any other phenomenon.

The first of these problems was by far the most difficult, and, indeed, its perfect solution remains yet to be accomplished, though, for any practical astronomical purpose, the problem has been solved in more than one way.

The plan adopted in the Cincinnati Observatory may be described as follows:—The clock-work machinery employed to give to the great equatorial telescope a uniform motion equal to that of the earth's rotation, on its axis, offered to me the first obvious approximate solution of the problem under consideration. This machinery was accordingly applied to the motion of the disk, or rather to *regulate* the motion of revolution, this motion being produced by a descending weight, after the fashion of an ordinary clock. It was soon discovered that the "Frauenhofer clock," as this machine is called, was not competent to produce a motion of such uniformity as was now required. Several modifications were made with a positive gain; but after long study it was finally discovered that when the machinery was brought into perfect adjustment, the dynamical equilibrium obtained was an equilibrium of instability; that is, if from a motion such as produced a revolution in one exact minute, it began to lose, this loss or decrement in velocity went on increasing, or if it commenced to gain, the increment went on increasing at each revolution of the disk. Now all these delicate changes could be watched with the most perfect certainty; as, in case the disk revolved uniformly once a minute, then the seconds' dots would fall in such a manner (as we shall see directly), that the dots of the same recorded seconds would radiate from the center of the disk in a straight line. Any deviation from this line would be marked with the utmost delicacy down to the thousandth of a second. By long and careful study, it was at length discovered, that to make any change in the velocity of the disk, to increase or decrease quickly its motion, in short, to restore the dynamical equilibrium, the winding key of the "Frauenhofer clock" was the point of the machinery where the extra helping force should be applied; and it was found that a person of ordinary intelligence, stationed at the disk, and with his fingers on this key, could, whenever he noticed a slight deviation from uniformity, at once, by slight assistance, restore the equilibrium, when the machine would perhaps continue its performance perfectly for several minutes, when again some slight acceleration or retardation might be required from the sentinel posted as an auxiliary.

The mechanical problem now demanding solution was very clearly announced. It was this: Required to construct an automaton which should take the place of the intelligent sentinel, watch the going of the disk, and instantly correct any acceleration or retardation. This, in fact, is the great problem in all efforts to secure uniform motion of rotation. This problem was resolved theoretically, in many ways, several of which methods were executed mechanically without success, as it was found that the machine stationed as a sentinel to regulate the going of the disk was too weak, and was itself carried off by its too powerful antagonist. The following method was, however, in the end, entirely successful. Upon the axis of the winding key, already mentioned, a toothed wheel was attached, the gearing being so adjusted that one revolution of this wheel should produce a whole number of revolutions of the disk. The circumference of this wheel was cut into a certain number of notches, so that, as it revolved, one of these notches would reach the highest point once in two seconds of time. By means of an electro-magnet a small cylinder or roller, at the extremity of a lever arm, was made to fall into the highest notch of the toothed wheel at the end of every two seconds. In case the disk was revolving exactly once a minute, the roller, driven by the sidereal clock, by means of an electro-magnet, fell to the bottom of the notch, and performed no service whatever; but, in case the disk began to slacken its velocity, then the roller fell on the retreating inclined face of the notch, and thus urged forward by a minute amount the laggard disk, while, on the contrary, should the variation from a uniform velocity present itself in an acceleration, then the roller struck on the advancing face of the notch, and thus tended slowly to restore the equilibrium. Let it be remembered that this delicate regulator has but a minute amount of service to perform. It is ever on guard, and detecting, as it does instantly, any disposition to change, at once applies its restoring power, and thus preserves an exceedingly near approach to exact uniformity of revolution. This regulator operates through all the wheel-work, and thus accomplishes a restoration by minute increments or decrements spread over many minutes of time.

With a uniformly revolving disk, stationary in position, we should accomplish exactly, and very perfectly, the record of one minute of time, presenting on the recording surface thirty dots at equal angular intervals on the circumference of a circle. To receive the *time dots* of the next minute on a circle of larger diameter, required either that the recording pen should change position, or that at the end of each revolution the disk itself should move away from the pen by a small amount. We chose to remove the disk. To accomplish accurately the change of position of the disk, at the end of each revolution, the entire machine was mounted on wheels on a small railway track, and by a very delicate mechanical arrangement accomplished its own change of position between the fifty-ninth and sixtieth second of every minute.

THE RECORDING PENS.—It now remains only to describe the simple machinery by which the clock records its beats, and the observer makes the record of his observation. These instruments are called the *recording pens*. That belonging to the clock is called the *time* pen; the one used by the observer the *observing* pen. They are constructed and operate in the following manner: A metallic arm is constructed with a short axis, perpendicular to its length. The extremities of this axis are pivots working in the jaws of a metallic frame, which supports the axis of the pen in a horizontal position. The longer arm of the pen reaches over into the center of the disk, and is armed at its extremity with a *steel point* or *styl s.* Upon the long arm of the pen and near the axis is located a piece of soft iron denominated an *armature*, and beneath this armature an electro-magnet is firmly fixed. This magnet is placed on the circuit closed by the *wire-cross* vibrating with the clock pendulum, and thus, at every dip of the cross into the mercury cup, the armature of the pen is suddenly drawn down on the head of the magnet, and the moment the circuit is broken a spring acting on the short arm of the pen lifts it from the head of the magnet. It is readily seen that in this way the *stylus* is brought down by a sudden shock or blow on the material placed on the revolving disk to receive the record. The pen is so adjusted that in case the armature be simply placed and held by hand on the head of the magnet, the steel point of the *stylus* does not quite touch the recording surface on the disk. The elasticity of the long arm of the pen is, therefore, a matter of the greatest moment, for this elasticity causes the pen to make a simple dot, by a sudden blow and recoil; whereas were the pen non-elastic, there would be a drag for the time during which the magnet holds the pen, which would at once destroy the uniformity in the going of the disk.

A pen constructed in precisely the same way, and placed at right angles to the former, so that the points of the two pens fall in close proximity on the disk, is operated by a magnet made by a circuit closed at will by the finger of the observer; and thus he is enabled to throw down upon the time scale a dot, which, falling between some *two-second dots* on the disk, records the exact instant of any phenomenon under observation.

When the disk is filled, we have only to lift it from its socket and replace it with a new disk. To read the time scale it is only necessary to mark on the disk from the clock face the time denoted by any one dot; for example, 12h. 15m. 00s. The circle next outside will be 12h. 16m., the next circle 12h. 17m., &c.; while the first or marked radius of dots will be the 0 second of all the minutes, the next in order will be the second, the next the fourth, and so on to the 58th and 0 second again. Thus we read the scale as rapidly as we read a clock face, for the hour, minute and second; and it only remains to construct a machine for measuring the fractions of seconds.

THE ANGULAR TIME MICROMETER.—This instrument is very simple. Take a common carpenter's two-foot rule; cut away the inner portion of one of the legs for two-thirds of its length, and insert a piece of plane glass; draw from the centre of the joint with a diamond point, on the under surface of this glass, a delicate straight line, and blacken by rubbing in black lead pencil. The arms of this micrometer are a little longer than the radius of the disk. To the left hand arm, at its outer extremity, attach a small brass arc, divided into seconds and tenths, and make it, say, 2½ seconds in length. When the two legs are closed the black line on the glass will read 0 on this scale of seconds. At the joint drill a small hole, and at the center of the disk to be measured erect a small vertical pin to fit this hole. Lay the instrument on the disk, the pin being inserted in the hole, and thus the fractions of seconds may be measured with any degree of precision.

Such is an outline of the machinery now in use in the Dudley Observatory at Albany, and at the Cincinnati Observatory.

As we have seen, in the old method of transits the attention of the observer was divided among many objects. He was compelled to keep up the counting of the clock beat; to estimate the space passed over by the star under observation in a second of time; to subdivide this space by estimation into tenths; to write down in his note-book the observed moment of transit across each of the wires, and all this while his eye continued to follow the movement of the object under observation. To give the observer time to make his record, the spider's lines or wires were necessarily separated by such an interval from each other that several seconds would be required by the star to pass from one to the other, and thus but few wires could be employed in transit observations.

In the new method the observer is released from all responsibility with reference to time, counting of clock beat, estimation of spaces, or entries in note-book. The clock records its own beat, and the observer has nothing to do but touch a magnetic key at the exact moment in which his star is bisected by the meridian wire. This touch records the moment of observed transit, and as this record is accomplished almost instantaneously, the observer is ready to record the transit across the next wire, and thus the interval between the wires may be greatly reduced, and their number extended almost indefinitely. While in the old method long practice was required to make an accomplished observer (the best of whom could not record more than the transits on seven wires), in the new method a few nights of practice gives all desirable experience, and the observer may record the transits across as many as fifty wires, should so large a number ever be desirable under any circumstances. It is found in the use of this method that erroneous habits of observation may either be entirely avoided or detected, and thus corrected. It furnishes the means of measuring with great accuracy the value of *personal equation*, and has demonstrated, indeed, that the large differences existing between observers, amounting in some instances to a whole second of time, are not due to physiological constitution, but almost entirely to false habits of observation. It has furnished the means of measuring the amount of time which elapses between the occurrence of any phenomenon falling within the grasp of the senses of sight and hearing, and the possible record by the touch of a magnetic key. In this operation there are three distinct processes, the sense of sight, for example, conveys to the brain information of the occurrence of the external phenomenon; the mind thus perceives, and the will issues an order to the nerves to record; the nerves execute this order. Thus far it has been impossible to ascertain the amount of time occupied in each of these processes, but the sum of the times, or that elapsing between the moment of occurrence of a phenomenon and its record, has been measured both for the sense of sight and the sense of hearing, in a large number of persons of both sexes and of all ages. From these experiments it has been ascertained that while different individuals present prominent and marked differences, these differences are only found to exist in the hundredths of a second of time, and not, as has been imagined, in whole seconds. In conducting these experiments it was ascertained that all observers, without a single exception, in attempting to mark the moment at which a star crossed a wire, *anticipated* the moment of transit, and the recorded time was thus in advance of the true time. Having learned this fact, the observer is placed upon his guard, and is furnished with the means of correcting this false habit, and of bringing himself up to a standard of positive accuracy. Another advantage derived from this mode of observation arises from the fact that it imposes but a slight tax upon the nervous system, and hence an observer is able to continue his work without exhaustion for a much longer period of time.

We have mentioned that one of the most hidden sources of error lies in the uncertainty of the *rate* of going of the clock. The old methods furnish the means of ascertaining with comparative accuracy how much the clock has lost or gained in twenty-four hours; and if this quantity should amount only to a fraction of a second, it is almost impossible to assert that this loss or gain may not have occurred even a hundred times, or possibly a thousand times during the twenty-four hours. By causing two or more clocks to record their beats upon the same time-scale, the new method furnishes the means of inter-comparison between these clocks, even from second to second, if required, and thus from a record of this kind may be obtained a positive standard of time.

The electro-magnetic method of observation in connection with the system of telegraphic wires, now extended over nearly all the civilized world, furnishes a very rapid and exact method of determining the *difference of longitude* between any two points. This difference of longitude is nothing more than the time which elapses from the transit of a star across the meridian of one place until it crosses the meridian of the other place. In case the two observatories whose difference of longitude is required are connected by telegraph, and are furnished with the electro-magnetic apparatus, the observer in the eastern observatory may send to his correspondent by telegraph the moment of transit of the star across his own meridian. He will receive in return by telegraph the moment the same star crosses the meridian of the western observatory, and in case the observations are perfectly made, transmitted with infinite velocity along the wires, and recorded with perfect accuracy, the result will be absolutely perfect. The common errors of observation are readily eliminated, the errors of recording, in like manner, are easily detected and measured, and the only matter of difficulty which remains is to ascertain whether the message sent along the wire travels at a *finite rate*, and if so, to determine *what* this rate may be. The conversion of time into space, and the delicacy of the machinery now employed in recording and subdividing time, has furnished the means of measuring the velocity with which signals are transmitted along the wires of the telegraph. No doubt this velocity is modified by a variety of circumstances, and may depend upon the direction in which the telegraphic wire is laid, the season of the year, the temperature of the earth and atmosphere, but none of these causes can interfere to mar the accuracy of the work employed for longitude purposes; for there is no difficulty in determining the exact velocity with which the signals are transmitted by the wires at the time of observation. These are a few among many advantages which have been gained by the conversion of time into space, and the application of this principle to the observation of astronomical transits.

The author has attempted to add something to the facility and accuracy of the determination of *north polar distances*, the second great element employed in fixing the place of a heavenly body. As already explained, this element is reached by the division of a circle attached to the axis of the transit, and the accuracy of the work depends upon the perfection of these divisions, the permanence of the figure of the circle, the permanence in the place of the reading microscopes, and the precision attainable in reading the subdivisions of the circle. As the errors which arise from these different sources are found to be comparatively large, for the measurement of small differences of north polar distances, or small arcs of space, resort has been had to other and more delicate mechanical contrivances, hence the invention and construction of the various *micrometers* now in use, all of which depend for their accuracy upon the performance of a *micrometer screw.* Very extended experiments with these instruments first created a doubt in my own mind as to the accuracy with which the micrometer screw would repeat its own measures. This doubt, added to the fact that the measurements by the micrometer were very slow and tedious, gave rise to the effort which has resulted in the construction of a new system whereby differences of north polar distance may be determined with great rapidity and precision, which principle can readily be extended to the determination of absolute north polar distances. A description of the machinery employed for this purpose may be found elsewhere. We are only concerned here to notice some of the possible advantages of this new method of north polar distances. I will only state that the machinery employed in all its joints and connections is of the simplest kind, and everywhere visible to the eye. There is no concealed portion, as in the screw micrometer, no joints to grow imperfect by wearing, and no strong resistance to change the figure of any part of the machinery. If the tube of the telescope, loaded as it is with the weight of the object glass and eye-piece, and its own weight, can be depended upon to retain its figure without a counterpoise, it is absolutely certain that the *declination arm*, which in the new method is attached to the axis of the transit, if perfectly counterpoised and bearing no weight whatever, can be relied upon to retain its figure. The lower extremity of this arm, moving as it does in north polar distance, with the line of collimation of the telescope, by a connecting bar, gives motion to the axis of the reading microscope, which, being directed to a distant scale, magnifies in a very high ratio by mechanical means the arc through which the transit revolves in the plane of the meridian. Thus it will be seen that this new method is nothing more than the use of a *mechanical magnifier*, and the only question is, can the scale be so divided as to read seconds of arc, and can it be made of invariable length? There is little difficulty in accomplishing both of these objects, for scales have already been divided with such precision that no error amounting to the *hundredth part of a single second of arc* could possibly exist; and in order to retain an invariable length in the scale all that is necessary is to grade it upon a surface constituting one face of a rectangular tube; fill this tube with water and broken ice, and thus a permanent temperature of 32° may be had for any length of time. To measure the exact value of the divisions on the scale we have only to employ these divisions in measuring around the entire circumference of the circle attached to the axis of the transit. Suppose the length of the scale to be sixty minutes approximately, then if this length is contained 360 times in the whole circumference, its approximate value becomes its absolute value, and at all events this experiment furnishes the means of determining the absolute value. Thus while the circle furnishes the means of measuring the *scale*, the scale furnishes in return the means of measuring the *subdivisions of the circle.* These amount only to 360, and may be reduced even to the fifth part of this number, should practice prove this reduction desirable. This small number of divisions can rapidly be read up with a scale of invariable length, and by performing this reading at temperatures widely different a correction for temperature may be determined with great exactness. In the old circle, as there are no less than ten thousand divisions, and as there exists no permanent scale for the reading of these divisions, it becomes almost impossible to learn their actual values and to tabulate their errors, hence astronomers have been compelled to rely to a great extent upon the assumed accuracy of the subdivisions of their circles, as received from the hands of the manufacturer.

By a combination of the electro-magnetic method, with the new method of measuring north polar distances, a very simple, convenient, and accurate instrument is obtained for recording the places of the stars or other heavenly bodies with great rapidity and exactitude, rendering it possible to construct, in a comparatively short time, a very extended and exact catalogue of the places of all the fixed stars, clearly visible, with any optical power.

We have thus presented a rapid sketch of the old and new methods of fixing the elements for the determination of the heavenly bodies, it only remains in this connection to speak of the optical power of the telescope.

These instruments are divided into two great classes, called *reflecting* and *refracting* telescopes. In the reflecting telescopes the rays of light from the external object, passing down the tube of the telescope, fall upon a metallic mirror or speculum, whose surface, perfectly polished, has the figure of a paraboloid of revolution, Being reflected by this surface, the rays of light are concentrated at a certain point, called the *focus*, where an intensely luminous image of the object is formed. This image is then examined by a magnifying glass or eye-piece, and its dimensions expanded to any required degree.

In the refracting telescope the light falls upon what is called the *object-glass*, a powerful lens, which concentrates, by refraction, the rays of light which pass through it, thus forming an image of the object at the focal point. This image is then examined, as in the reflecting telescope, by eye-pieces having different magnifying powers. Hitherto it has been found impracticable to construct *object-glasses* of any very considerable diameter, the largest of these glasses in use not exceeding sixteen to twenty inches in diameter. These narrow limits do not exist, however, in the construction of the *metallic specula* which belong to the reflecting telescope; and hence we find gigantic instruments have been constructed by different observers, one of which, now in use by Lord Ross, has a speculum of no less than *six feet* in diameter, with a focal length of fifty-two feet. Such immense instruments, requiring ponderous machinery for their management, are not well adapted for that kind of observation having for its object to determine the *places* of the heavenly bodies. Their use has been rather confined to examinations of the planets, double stars, clusters, and nebulæ, demanding a large amount of light rather than a perfect definition or exactitude in measurement. It is true, that in the hands of Lassell, of Liverpool, we find the reflecting telescope performing admirably in the routine work of an observatory. But these instruments are comparatively small, their dimensions not much exceeding those of the largest refractors.

There are two qualities which distinguish the telescope, *the space-penetrating power* and *the power of definition.* The first of these depends exclusively upon the amount of light received and refracted, or reflected to the focus, and thus forming the image. In case all the light falling upon the object-glass or speculum could be concentrated in the formation of the image, then the space-penetrating power of telescopes would be exactly proportioned to the diameters of their apertures, and we can compare then, readily, the space-penetrating power of different instruments, not only among themselves, but directly with the space-penetrating power of the human eye. The diameter of the pupil of the eye determines the amount of light which can enter and form the image, just as the diameter of an object-glass in a telescope determines the amount of light which in that instrument forms the focal image; hence, if we desire to know how many times deeper a telescope can penetrate space than the eye, we have only to learn how many times the area of the object-glass exceeds that of the pupil of the eye. We shall have occasion hereafter to employ this principle when we come to examine the relative distances to which the nebulæ and clusters are sunk in space.

We have only spoken of the mounting of the transit, with its attached circle, for reading north polar distances. This instrument revolves only, as we have seen, in the plane of the meridian, and of course no object can be seen with the transit except when in the act of passing the meridian line.

A telescope mounted in such a manner that it can be directed to *any point* of the heavens, is called an *extra-meridional instrument*, and of these the *equatorial* is the most used, and is the best adapted for all observations off the meridian. The tube of the telescope is carried by a heavy metallic casting, very firm and strong, which is made fast to a metallic cylinder, through which passes a steel axis, called the *equatorial axis*. The metallic cylinder is also screw-bolted to the extremity of a heavy steel axis, so placed on its supports as to lie parallel to the earth's axis. These supports rest on heavy metallic plates, bolted to a massive stone pier, called the "foot of the instrument," which, in turn, is placed on the top of a heavy pier of masonry, resting on a rock foundation, or something equally solid, and entirely disconnected from the building.

The instrument is so counterpoised in all its many parts as to be readily moved either on its polar or equatorial axis, and may thus be directed to any point of the celestial sphere. To enable the observer to follow the object under examination these telescopes are usually furnished with a species of *clock-work*, which causes the instrument to revolve round its polar axis with a velocity equal to that of the earth's rotation, causing it to follow a heavenly body and to hold it steady in the field of view for any required period of time.

Without extending further our notice of the instruments employed in reaching the data required in astronomical investigation, we will now return to our examination of the bodies which compose the sun's retinue, and shall proceed in our plan, preserving the order of distance from the sun.

The interruption which was made after closing the

discussion of the system of Saturn, to introduce to the student the laws of motion and gravitation, and the instruments employed in astronomical measures, was necessary to a full comprehension of the extraordinary investigations which are now to follow. We are hereafter to treat the planets and their satellites as ponderable bodies, mutually affecting each other, and all subjected to the dominion of the laws of motion and gravitation.

CHAPTER XII.

URANUS, THE EIGHTH PLANET IN THE ORDER OF THE DISTANCE FROM THE SUN.

ACCIDENTALLY DISCOVERED BY SIR WILLIAM HERSCHELL.—ANNOUNCED AS A COMET.—ITS ORBIT PROVED IT TO BE A SUPERIOR PLANET.—THE ELEMENTS OF ITS ORBIT OBTAINED.—ARC OF RETROGRADATION.—PERIOD OF REVOLUTION.—FIGURE OF THE PLANET.—INCLINATION OF ITS ORBIT.—SIX SATELLITES ANNOUNCED BY THE ELDER HERSCHELL.—FOUR OF THESE NOW RECOGNIZED.—THEIR ORBITAL PLANES AND DIRECTIONS OF REVOLUTION ANOMALOUS.—EFFORTS MADE TO TABULATE THE PLACES OF URANUS UNSUCCESSFUL.—THIS LEADS TO THE DISCOVERY OF A NEW EXTERIOR PLANET.

IT was remarked at the close of our investigation of the Saturnian system that this planet inclosed by its orbit all the objects belonging to the solar system which were known to the ancients, and whose phenomena, as observed and recorded in all time, furnished the data for the discovery of Kepler's laws and the law of universal gravitation, as finally revealed by Newton. While many of the modern astronomers, from an examination of the inter-planetary spaces, had ventured to suggest the probable existence of a large planet revolving in an orbit intermediate between those of Mars and Jupiter, no one had ventured to predict the possible discovery of planets lying exterior to the mighty orbit of Saturn. From the very dawn of astronomy this planet had held the position of sentinel on the outposts of the planetary system, and many strong minds had long entertained the opinion that no other bodies existed exterior to the orbit of Saturn forming a part of the scheme of worlds revolving around the sun. Such, indeed, was the prevalence of this opinion that when, in 1781, Sir William Herschell, in a course of systematic exploration of the heavens, discovered an object having a well-defined planetary disk, and whose movement among the fixed stars became measurable, even at the end of a few hours, he did not even suspect this new object to be a planet, but announced to the world that he had discovered a most extraordinary *comet*, without any of the usual haziness which attends these bodies, but presenting a clear and well defined planetary disk.

This newly-discovered object soon attracted universal attention. It was observed at the royal observatory at Greenwich, and the then astronomer royal, Dr. Markelyne, was the first to suspect its planetary character. Efforts were made by several computers to give to the new comet, as it was called, a parabolic orbit; this, however, was found to be impossible, and it was very soon found that the newly-discovered object was revolving around the sun in an orbit nearly circular in form, lying in a plane, nearly coincident with the ecliptic, and completing its mighty revolution in a period of no less than *eighty-two* years. It must be remembered that these extraordinary discoveries and announcements were made at the end of a very short examination, while the periods of revolution of all the old planets had been obtained from actual observation, through long centuries of patient watching. The periodic time of this last discovered of all the planets, which, by the old method of watching its return to the same fixed star, could not have been determined in less than eighty-two years, and even then only approximately, was, by the new method, based upon the law of universal gravitation, guided by the results of a few nights of accurate observation, and worked out by the powerful formulæ of analytic reasoning, given to the world with accuracy after only a few months of investigation. This is the first illustration of the change wrought in the whole movement of astronomical science by the great discoveries of Newton, and by the almost equally extraordinary step accomplished by Descartes, in fastening the powers of analysis upon geometry. All the circumstances of motion of this planet were rapidly investigated; the eccentricity of its orbit; the position of its perihelion; the inclination of its orbit to the plane of the ecliptic; the position of its line of nodes; the measure of its actual diameter; the determination of its various distances from the sun, all these and many other peculiarities were accurately determined from actual observation and computation. These facts strike us with the more astonishment when we reflect that the planet Uranus is removed to a distance of *eighteen hundred millions* of miles from the sun, and that, although its actual diameter is *thirty-five* thousand miles, it is absolutely invisible to the naked eyed, and, when seen through the most powerful telescope, presents a disk of only the *five hundredth* part of the apparent diameter of the sun. At such an immense distance it has been impossible thus far to determine anything with reference to the precise figure of Uranus. The discoverer of the planet thought that he saw a flattening at the poles, but subsequent observation has not confirmed this announcement. We have only, therefore, analogy to induce us to believe that this planet, like all the others, rotates upon an axis, and that, consequently, its figure is that of the elipsoid and not of the sphere. The immense magnitude of the orbit of Uranus, when compared with that of the earth, causes this planet to retrograde over an arc of only 3° 36′, but the duration of the retrograde motion extends over a period of no less than one hundred and fifty-one days. No telescope has yet been able to discern, upon the surface of Uranus, any spot or belt, or any well-defined point, distinguished from the entire surface, so that we have no means, thus far, of fixing the period of rotation upon its axis. The amount of light and heat received by Uranus, admitting the law of diminution, which seems to govern these elements, could only be the quarter part of that received by the planet Saturn, while the apparent diameter of the sun, as seen from Uranus, would be less than the thirtieth part of his diameter, as seen from the earth.

This planet is surrounded by at least four satellites. Two others were announced by Sir Wm. Herschel, who not only gave their distances but their periods of revolution, yet no telescope has since been able to detect these minute points of light, and their very existence is now doubted by many of the best observers. Four of the satellites had been studied, with much care, and their periods of revolution and their mean distances had been well determined. Of these, the second and fourth are most readily seen, and different astronomers have obtained results which agree with each other within comparatively small limits of error. Thus, the elder Herschel fixed the period of revolution of the second satellite, in the order of distance, at 8d. 16h. 56m. 5s. Sir John Herschel made the same period twenty-six seconds longer. Dr. Lamont, of Munich, obtained, for this same period, a value of 8d. 16h. 56m. 28s.5. The period of revolution of the fourth satellite, in the order of distance, as determined by Lamont, amounts to 13d. 11h. 07m. 06s.3. The period of revolution of the nearest satellite is about five days and twenty-one hours, while the third satellite in order of distance performs its revolution in a period of about eleven days. These are among the most difficult of all the objects revealed to the eye by telescopic power. After Sir Wm. Herschell no one for many years was able to see any of these satellites, the forty-foot reflector of Herschel having gone into disuse. In 1828, Sir John Herschel, after many unsuccessful attempts, by confining himself in a dark room for many minutes previous to observation, and thus giving to the eye great acuteness, succeeded in detecting two of these satellites. In 1837, Lamont, with the powerful refractor of the royal observatory of Munich, managed to follow, with tolerable certainty, the two larger satellites, and occasionally obtained glimpses of two others.

At this time there are four or five telescopes in the world capable of showing these four satellites, under favorable circumstances. I have frequently seen two of them with the Cincinnati refractor, but they are certainly objects of great difficulty, and only to be discerned under the most favorable circumstances in the observer, and under the best possible conditions of atmosphere.

Enough, however, has been determined with reference to these four satellites to warrant the assertion of a fact of most extraordinary character, and nowhere else to be found in the whole range of the solar system, namely, *that their orbits are nearly perpendicular to the plane of the ecliptic, and that their motions are retrograde.* We have seen that all the planets revolve in orbits whose planes are nearly coincident with the plane of the ecliptic; that they all revolve in the same direction around the sun; that the sun and all the planets rotate on their axes in the same direction in which they revolve in their orbits. We have found, in like manner, that all the satellites of every planet revolve around their primaries in the same direction, and in planes nearly coincident with the planes of the equators of their primaries; so that it became a settled opinion that there was but one direction in which any rotation or revolution could be performed by a member of the planetary system; and thus when the asteroids were discovered, although there were considerable deviations in the angles of the inclination of the planes of their orbits from those of the old planets, yet in every instance their motions are found to be direct. These satellites of Uranus present, then, the only example of retrograde movement among the legitimate members of the solar system. We shall see hereafter that among the comets (which may be regarded as satellites of the sun) there are a few which present this same anomaly of retrograde movement, yet this is not nearly so surprising as to find this anomalous motion among the satellites of a primary planet. We shall return to the consideration of this subject when we come to discuss the cosmogony of the universe.

If we recall to mind the relations which exist between the distances and periodic times of Uranus and Saturn, we shall find that these two planets, when nearest to each other, or when in conjunction, are separated by a distance of about nine hundred millions of miles. When most remote from each other, this distance of separation is increased by the whole diameter of the orbit of Saturn, or by eighteen hundred millions of miles, as will be readily seen from the figure, in which S represents the sun,

A and B the places of Saturn and Uranus when in conjunction, while B′ represents the place of Uranus in the opposite part of its orbit, or when in opposition to the sun. Thus the distance between the planets when located at

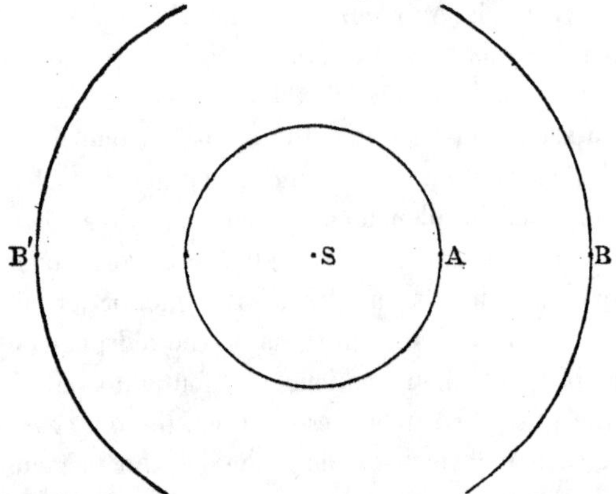

A and B is just equal to the interval between their orbits, while this interval is increased as Uranus recedes from B up to the time that it reaches B′, and on reaching this point, Saturn being supposed to occupy the point A, the two planets will be separated by a distance of about twenty-seven hundred millions of miles. Since Saturn performs its revolution in about twenty-nine years and a half, and Uranus performs its revolution in about eighty-two years, the interval from one conjunction to the next is readily computed to be about forty years.

This extraordinary change of distance produces a corresponding change in the reciprocal influences exerted by these planets upon each other. The same remark is applicable to the configurations of Jupiter and Uranus, and may be extended indeed to all the planets. Thus we perceive that the greatest possible effect to draw Uranus closer to the sun will be produced when all the planets lie on the same straight line, and on the same side of the sun.

The prevalence of the law of universal gravitation, whereby every particle of matter in the universe feels the attraction of every other particle, unites all the planets and their satellites into one grand scheme of revolving worlds, in which each is subjected to the influence of all the others. After the discovery of Uranus an effort was made to assign to this planet a curve whose magnitude and position were derived from observations embracing but a small portion of its orbit. This, of course, was a matter of necessity, for even one revolution has not yet been completed by Uranus since the date of its discovery in 1781. The orbit assigned to the planet was sufficiently accurate to trace backward its movement among the fixed stars. This was done in the hope that the planet might have been seen and its place recorded as a fixed star by some of the early astronomers. If it should happen that the computed place of the planet should coincide with the recorded place of a star of the same magnitude as the planet, then a suspicion would arise that this star and the planet were one and the same body. If on directing the telescope to the point once occupied by the star the place should be found vacant, this evidence would be almost conclusive that the supposed star was actually the planet. By an examination of this kind it was found that the planet Uranus had been observed, and its place carefully recorded by no less than three astronomers, each of whom had seen it several times, without any suspicion of its planetary character. The astronomer Flamsteed was the first who had mistaken this planet for a star nearly ninety years before its discovery by Sir William Herschell. It was subsequently observed by Bradley, by Mayer, and by Le Monnier, who fixed its place no less than twelve times during the period from 1750 to 1771. These ancient observations furnished an opportunity to test the accuracy of the computed elements of the orbit of the new planet, and to correct these elements in case they were found to be sensibly in error. This work was executed in a most faithful and exact manner by M. Bouvard, who also computed tables predicting the places of Uranus for many years in advance. It was supposed with reason that these tables would point out the places of Uranus with the same certainty as those of Saturn and Jupiter, computed by the same astronomer, gave the places of these planets. In this the hopes of the astronomical world were disappointed, and this extraordinary discrepancy between computation and observation gave rise to the discovery of an exterior planet, as we shall now relate.

CHAPTER XIII.

NEPTUNE, THE NINTH AND LAST KNOWN PLANET IN THE ORDER OF DISTANCE FROM THE SUN.

Uranus Discovered by Accident.—Ceres by Research with the Telescope. —Rediscovered by Mathematical Computation.—The Perturbations of Uranus.—Not Due to any known Cause.—Assumed to Arise from an Exterior Planet.—Nature of the Examination to find the Unknown Planet.—Undertaken at the same time by two Computers.—Computation Assigns a Place to the Unknown Planet.—Discovered by the Telescope.—Discoveries Resulting.—A Satellite Detected.—The Mass of Neptune thus Determined.—Neptune's Orbit the Circumscribing Boundary of the Planetary System.

The discovery of Neptune is undoubtedly the most remarkable event in the history of astronomical science—an event without a parallel, and rising in grandeur pre-eminently above all other efforts of human genius ever put forth in the examination of the physical universe.

The planet Uranus was discovered by the aid of the telescope, not exactly by accident, but still without any expectation on the part of the discoverer that his examination of the fixed stars would result in the addition of a primary planet to the system. Indeed, as we have seen, so little did the astronomical world then anticipate the discovery of a new planet that the announcement by Sir William Herschel that he had detected a most remarkable comet was accepted on all hands, and it was only continued observation that finally compelled astronomers to accept the new object as a planet. In the case of the discovery of the first asteroid we find a systematic organization of astronomical effort to detect a body whose existence was *conjectured,* on the single ground of the harmony of the universe, or that the law of interplanetary spaces, interrupted between Mars and Jupiter, would be restored by finding a planet revolving within that vast interval. Hence a search was commenced which consisted in examining every star in the region of the ecliptic, to ascertain whether its place was already laid down on any known map or chart of the heavens. Now it is evident that if it were possible to make a perfect daguerreotype of any region of the celestial sphere, say to-night, and the same could be effected on the following night, the comparison of these two pictures would exhibit to the eye any change which may have occurred in the interval from the one picture to the other; and hence if a star was found on the second and not on the first picture, this star might fairly be suspected to be a planet, or the same suspicion would attach to a star found on the first, but missing on the second picture. Now, a map of the heavens, so far as it includes the correct places of the stars, answers our purpose quite as well as the daguerreotype, and any star found in a region well charted, but not laid down on the map, may be fairly suspected to be a planet. A few hours of examination will show it to be at rest or in motion. If in motion, then its planetary character is decided.

This method of research has been employed in the discovery of all the asteroids, and there is but one example in which a more powerful and searching examination became necessary. This was in the case of the asteroid Ceres, which, as we have seen, was discovered by Piazzi, at a time when but few observations could be made previous to its being lost in the rays of the sun. For a long time it seemed almost a hopeless task to undertake the re-discovery of the planet, as the telescope would be compelled to grope its way slowly round the heavens, in the region of the ecliptic, comparing every star with its place in the chart. In this dilemma mathematical analysis essayed to erect a structure on the narrow basis of the few observations obtained by Piazzi, whereon the instrumental astronomer might stand and point his telescope to the precise point occupied by the lost planet. The genius of Gauss succeeded in this herculean task, and when the telescope was pointed to the heavens in the exact place indicated by the daring computor, there, in the field of view, shone the delicate and beautiful light of the long-lost planet.

This was certainly a most wonderful triumph of analytic reasoning, yet in this case the planet had been discovered, was known to exist, and had been observed over 4° out of the 360° of its revolution round the sun. On this basis of 4° it was *possible* to rise to a knowledge of the planet's position at the end of a few months of time.

The case of the discovery of Neptune is entirely different. Here no planet was known to exist, no telescopic power, however great, had ever seen it. For ages it had revolved round the sun in its vast orbit, far beyond the utmost known verge of the planetary system, unfathomably buried from human gaze and from human knowledge. No sage of antiquity had ever dreamed of its existence. The fertile brain of even Kepler had failed to imagine its being, and the powerful penetration of Newton's gigantic intellect had failed to pierce to the far off region inhabited by this unknown and solitary planet.

Indeed, with the knowledge which existed prior to the discovery of Uranus, no human genius, however mighty, could have passed the tremendous interval which separates the orbits of Saturn and Neptune from each other. The discovery of an intermediate planet was requisite to furnish a firm foothold to him who would adventure to pass a gulf of not less than 2,000 millions of miles at its narrowest place.

We shall now proceed to relate the circumstances which led to the discovery of Neptune. As already stated, a careful and elaborate study of the orbit of Uranus had been accomplished by M. Bouvard, and tables giving the computed places of this planet had been prepared by the same astronomer. It was not anticipated that these tables would be absolutely perfect, even if based on perfect observations. We must remember that each body of the solar system affects every other, and hence no single set of observations are sufficient to give a perfect orbit. In case all the other worlds were blotted out of existence, and there remained only the sun and Uranus, then three perfect observations of the planet's place would suffice to determine positively all the elements of its orbit, and fix forever all the circumstances of its motion. We shall call the figure of the orbit of Uranus, obtained under the above hypothesis, the *normal figure*, and the ellipse which it would describe about the sun, under the above circumstances, the *normal ellipse*. If now we introduce another planet into our system, as, for example, Saturn, it is possible, as we have already seen, to compute the exact amount of power exerted by Saturn to disturb the movements of Uranus, and to change the figure of its orbit. In like manner, by adding successively all the interior planets, it is possible to compute the perturbations that each produces upon the orbit of any particular one, until, finally, by using all the power of analytic reasoning, the human mind may reach to a complete knowledge of all possible derangements produced by the combined action of all existing known causes of perturbation.

Supposing our knowledge in this way to become perfect as to the movements and orbit of Uranus, we can then predict its places in all coming time, and these predictions, being arranged in tabular form, may be verified by comparison in after years with the observed places of the planet. If now a new planet were added to the system, revolving in an orbit exterior to that of Uranus, perturbation would arise from the introduction of this disturbing body into our system which would at once cause the planet to deviate from its predicted track, and the observed and computed places would no longer agree.

We can perceive at once, from this statement of the problem, that these very discrepancies between the old track and the new one, pursued by the planet, would give us a clue whereby it might become possible to determine, in space, the position of the disturbing body.

Difficult and incomprehensible as the above problem may appear, it is far less difficult than the one actually presented in nature. We have supposed the normal ellipse, described by Uranus, to be known, whereas, in reality, this very ellipse had to be determined by a train of reasoning of the most searching and powerful character, while the whole problem was almost hopelessly embarassed by the fact that the movements of Uranus were actually being disturbed all the time by the unknown body whose position in space was required. As the normal orbit could only be determined by a series of approximations, based upon the observed places of the planet, it was impossible, in any one of these approximations, to free Uranus from the disturbing effects of the unknown body. It was only, therefore, by comparing with each other the results reached by these successive approximations to the orbit of Uranus that it became manifest that no increase of accuracy was being reached by these successive efforts, and after every known cause of disturbance had been carefully taken into account, a grand conclusion was finally reached that no satisfactory account could be rendered of the movements of Uranus by the combined effects of all *known* disturbing causes.

To reach this conclusion required investigations of the most profound and laborious character, but before it was possible to explain these anomalous movements of Uranus, a problem of far greater difficulty remained to be solved, involving nothing less than a determination of the *weight* of the unknown planet, its *distance* from the sun, the *nature of its orbit*, and its *position* in the heavens at a particular time, indicating the region to which the telescope must be pointed to render visible what had hitherto remained for ages unseen by the eye of man.

To those who have given but little attention to the study of these extraordinary problems an attempt to resolve a question like that just presented may seem to be even presumptuous, yet when we come to examine the circumstances, we shall see dimly a way whereby we may reach a certain approximate knowledge of the place of this unknown world.

The fact that the planes of the orbits of all the more distant planets are nearly coincident with the ecliptic reduced the examination to the great circle in the heavens cut out by the indefinite extension of this plane. This is a most important consideration, and but for this fortunate circumstance no powers of research could ever have made even the most distant approximation to the place of the unknown planet.

The empirical law of Bode, whereby it seemed that the order of distances of the planets was governed, assigned to the hypothetical world a distance about double that of Uranus, or say, 3,600 millions of miles from the sun. Assuming this as the probable distance, the third of Kepler's laws would determine the period of revolution of the world whose position was sought. It remained now to assign a mass and position to the planet such as would render a satisfactory account of the perturbations of Uranus, which remained outstanding after the known causes of disturbance were exhausted. To accomplish this let us recall to mind the fact that in case the mean distance of the disturbing body had been rightly selected, then the interval between Uranus and the *unknown*, when in conjunction, would be about 1,800 millions of miles. In this position the disturbing force would exhibit its maximum power in a twofold sense: first, to cause Uranus to recede to its greatest distance from the sun; and, second, to cause the same planet to lag behind the place it would otherwise have reached. In the above figure let U U′ U″ represent the computed orbit of Uranus, as existing under the combined influence of all known causes, P the place of the *unknown*, when in conjunction with Uranus at U′. It is manifest that the force exerted by P on Uranus will tend to accelerate its velocity in coming up to conjunction, and to cause the path described to lie outside the computed path along the dotted line, the planet really reaching the point U‴, instead of U′ in the undisturbed orbit. Leaving this point the force exerted by the *unknown* would reverse its effect, and a retardation would commence, and by slow

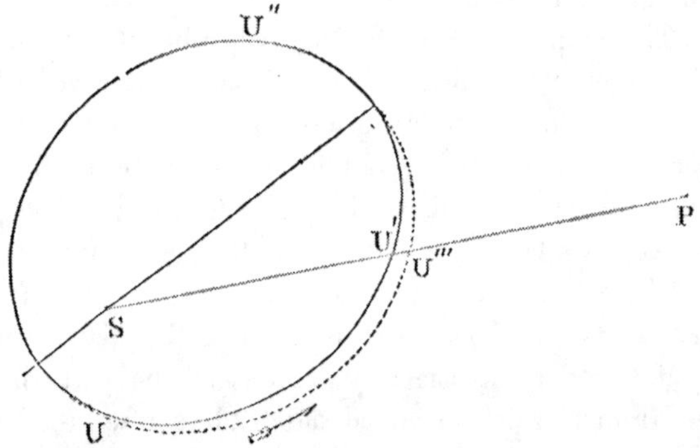

degrees receding from the disturbing body, it would gradually return to the undisturbed orbit, and there continue until the period for the next conjunction might approach.

Such is a rough exhibition of the reasoning which was employed to narrow the limits of research in the effort to point the telescope to the unknown cause of the perturbations of Uranus. No account, of course, can be given of the mathematical treatment of the problem. It was undertaken at about the same time by Adams, of England, and by Le Verrier, of Paris. Each computer, unknown to the other, reached a result almost identical. Le Verrier communicated his solution to the Academy of Sciences on the 31st August, 1847, and on the evening of the 18th September, 1847, M. Galle, of Berlin directed the telescope to the point in which the French geometer declared the unknown planet would be found

A star of the eighth magnitude appeared in the field of view, whose place was not laid down on any known chart. Suspicion was at once aroused that this might possibly be the planet of computation, and yet it seemed incredible that a problem far surpassing in difficulty any which had ever been attempted by human genius should thus at the first effort have been solved with such marvelous precision.

The suspected star was examined with the deepest interest in the hope that it might exhibit a planetary disk. In this, however, the astronomer was unsuccessful, and there remained but one method by which its planetary character might be determined, that of watching sufficiently long to detect its motion. This process, however, must have tried very sorely the patience of the observer, as the motion of the planet at so great a distance as three thousand six hundred millions of miles, was so slow as to require three entire months to pass over a space equal to the apparent diameter of the moon. The position of the suspected star having been accurately determined on the first night of observation, it became evident on the next night that the star had moved by an amount such as was fairly due to the slow motion of so vast an orbit. It could be none other than the unknown planet! A success almost infinitely beyond the expectations of the most sanguine computer had crowned this mighty effort, and the amazing intelligence that the planet was found startled the astronomical world.

The planet was soon recognized by the astronomers in every part of the world. The elements assigned by Le Verrier and Adams by computation were accepted everywhere with most unhesitating faith in their accuracy, and it was believed that it only remained for the telescope to verify the computations of these most wonderful mathematicians. In this the astronomical world were destined to meet a most remarkable disappointment. The new planet proved, indeed, adequate to account for all the anomalous movements of Uranus, while in all its elements it differed so widely from those of the computed hypothetical planet that the computed and real planet could not in any way be regarded as the same body. The first restriction proved to be correct, for the orbit of the new planet (afterwards named Neptune) did coincide almost exactly with the plane of the ecliptic. The second restriction, based on the extension of Bode's law of interplanetary spaces, was falsified in the event, for here the law of Bode failed, and the distance of the true planet was nearly 5,000 millions of miles less than that of the computed one.

The third restriction due to the application of Kepler's third law is verified in the real planet; but as the distance of the unknown was assumed greatly too large, of course the periodic time depending on this distance was also too large. This by necessity involved an error in the mass assumed for the unknown, whose erroneous distance demanded, of course, an erroneous mass greater than that of the true planet; and yet, notwithstanding the magnitude of the errors of these elements, the computers succeeded in pointing the telescope within less than one degree of the actual place of the body which had caused the anomalous movements of Uranus!

We will endeavor to render a brief account of this most astonishing fact. It is evident that the disturbing effects of Neptune will become most powerful when the disturbing planet is nearest the disturbed one, or, what amounts to the same thing, the maximum disturbance will occur when the planets are in *conjunction.* We know that the periodic time of Uranus is 82 years, the periodic time of Neptune is 164 years, and hence it is easy to compute the interval from one conjunction to the next, which is no less than 171 years. The two planets passed their conjunction in 1822, and therefore the previous conjunction must have occurred 171 years before, or in 1651; but the earliest recorded observation of Uranus was not made till 1690, or nearly 40 years after the conjunction, and at a time when the disturbing force of Neptune was so much diminished as to be nearly, if not quite insensible, for a long while. The minute disturbing power of Neptune still existing in 1690, would go on decreasing until, in 1732, the planets would be in opposition, and would be separated by a maximum interval. After this date the distance between the planets would slowly decrease as they approached their conjunction, and in 1781, when Uranus was discovered, a small disturbing effect would begin to be appreciable, which would go on increasing up to the time of conjunction in 1822. Thus we perceive that mathematicians found the planet Uranus in such condition that the perturbing effects of Neptune were increasing in intensity from year to year; and hence no set of elements could correctly represent the places of Uranus, because the observations did not extend back far enough to embrace the disturbed places of the planet at the former conjunction in 1651. No correct solution was then possible until the perturbations should reach their maximum value, which occurred in 1822, when the planets were in conjunction, and subsequent to which period the planet Uranus would slowly return to its computed orbit as it receded further and still further from the influence of the disturbing body, as may be more clearly seen from the figure below, in which the smaller circle may represent the orbit of Uranus, the larger one the orbit of Neptune. For a long while prior

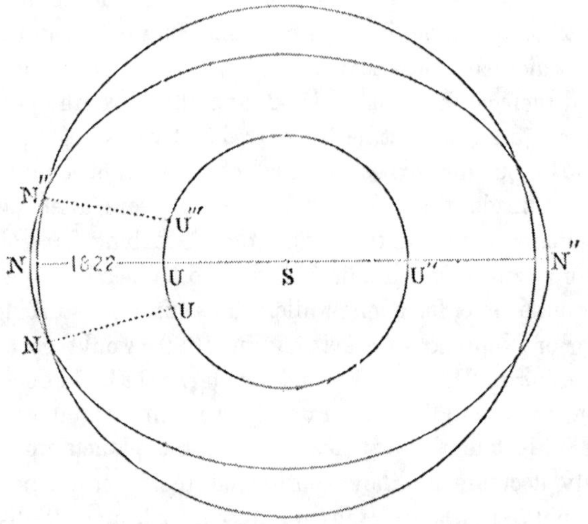

to conjunction in 1822, Uranus would be slowly overtaking Neptune, during which time the direction of the disturbing force would be such as to accelerate the orbital motion of Uranus, and to increase its distance from the sun. The acceleration would cease at conjunction, and would there be changed into equal and opposite retardation, as is manifest from the figure, while the increase of the distance of Uranus must continue to increase even after conjunction, but the disturbing force must rapidly decline in power as the interval between the planets increases. Thus the great problem demanded the position of a disturbing planet at a given time, which could account for the known perturbations, all of which were crowded into a few years, say 25, before and after the conjunction in 1822. While this narrowing of the limits of sensible perturbation increased the chances of directing the telescope to the unknown disturber, it seems to have really increased the difficulty of assigning to this disturber his exact orbit. Indeed, even with circular orbits, several might have been chosen, such that by varying the mass of the unknown the perturbations might have been tolerably well represented, but in case elliptical orbits are chosen, then our limits are much extended, and the mean distance may be made to vary within very broad limits, provided the eccentricity may be chosen at pleasure. Thus the ellipse shown in the figure coincides between N and N′ very nearly with the circular orbit, and in case a planet revolving in the circle could account for the anomalies of Uranus, the same would be tolerably well represented by the effects of a planet with a very different period and mean distance revolving in the elliptic orbit. Now this was exactly the case as developed in the final history of this grand discovery. The great geometers chose an elliptic orbit of such eccentricity and having its major axis in such position that the computed and true orbits agreed with each other in a most remarkable manner during the twenty years before and after conjunction. Their efforts were thus crowned with the success which they so eminently deserved; and although the computed orbit came finally to differ greatly from the true one, yet, for the time when the computed orbit was required to represent the places of the unknown, and to point the telescope to its actual location, the computed orbit responded nearly as perfectly as the true one could have done, even had it been then known.

It has been already stated that after the discovery of

Uranus, when the elements of its orbit had been obtained with sufficient accuracy to render it possible to trace the planet backwards among the fixed stars, it was ascertained that it had been observed and its place recorded as early as 1690, and had been seen many times subsequently and prior to its discovery, being always mistaken for a fixed star, so we find in the case of Neptune, a like examination by Professor Walker led to the discovery that the new planet had been twice recorded in position by La Lande, in May, 1795. These two observations were found to be outside the path which had been assigned the planet by the theory of both Le Verrier and Adams; and such was the deep confidence in the accuracy of the elements assigned by these two geometers that it was with great difficulty that some of the ablest astronomers could be induced to believe that the missing star twice observed by La Lande could be the new planet. The identity was, however, soon demonstrated, and hence arose the discussion which led to the declaration by an eminent mathematician that the discovery of Neptune was the result of a *happy accident;* but we have seen that the grand problem propounded by both the French and English astronomer, and which each resolved with such astonishing precision, was to point the telescope in the direction of the unknown, which had produced the late excessive perturbations of Uranus. It remains, so far as I know, yet to be decided whether the data in possession of Adams and Le Verrier can be so treated by analysis as to give an orbit to the unknown more nearly agreeing with that of the known planet.

NEPTUNE'S SATELLITE.—The vast distance to which Neptune is buried in space will perhaps render it impossible to learn how many satellites revolve about this remote primary. The great refractors have certainly discovered the existence of one satellite, and another is suspected. The discovery of this one satellite of Neptune becomes, under all the circumstances, a matter of deep interest, as it enables us to determine the mass or weight of the primary, a matter of the first moment in computing the effects of the planet as a disturbing body. The satellite is found to perform its revolution about the primary in a period of about five days and twenty-one hours, and at a mean distance of 232 thousand miles, or nearly equal to the distance of our moon from the earth. In case these distances are assumed to be exactly equal, then as at the same distance the centrifugal force increases as the *square* of the velocity, and as the velocity of Neptune's moon is about four and a half times greater than that of our moon, its centrifugal force in its orbit must be 4.5×4.5, equal to about twenty times the centrifugal force of the moon. Now, the attractive force of Neptune is exactly proportioned to its weight or mass, and hence, to counterbalance this centrifugal force in his satellite, which is twenty times as great as that of the moon, the mass of Neptune must be twenty times as great as that of the earth. Thus has been revealed not one world, but two—the one containing a mass of matter sufficient to form no less than twenty worlds as heavy as our earth—the other a satellite, indeed, of the first, yet sufficiently large to send back to us, at a distance of 3,000 millions of miles, the light of the sun, enfeebled by its dispersion over this vast distance to the one-thousandth part of the intensity it pours on our earth. We have reached the known boundary of that mighty confederation of revolving orbs which, whilst they acknowledge in the most specific manner a mutual dependence, are all controlled by the predominating influence of the sun, which occupies the common focus of all their orbits, and around which they all roll and shine in obedience to the grand law of universal gravitation.

We shall now retrace our steps toward the sun, and consider a remarkable class of bodies, which for ages were regarded as evanescent meteors, suddenly blazing athwart the sky, and as suddenly fading from the vision, never more to reappear. Modern science has given to these bodies determinate orbits, and in some instances, as we shall see, has assigned them a permanent place among the satellites of the sun.

CHAPTER XIV.

THE COMETS.

Objects of Dread in the Early Ages.—Comets Obey the Law of Gravitation and Revolve in some one of the Conic Sections.—Characteristics of these Curves.—Comet of 1680 Studied by Newton.—Comet of 1682 named "Halley's Comet."—Its History.—Its Return Predicted. Perihelion Passage Computed.—Passes its Perihelion 13th April, 1759.—Elements of its Orbit.—Physical Constitution.—Nucleus.—Envelopes.—Tail.—Intense Heat Suffered by some Comets in Perihelio. —Dissipation of the Cometic Matter.—Encke's Comet.—A Resisting Medium.—Deductions from Observation.—Biela's Comet.—Divided.—Number of Comets.

In all ages of the world these anomalous objects have excited the deepest interest, not only among philosophers, but among all classes of men. The suddenness with which they sometimes blaze in the sky, the vast dimensions of their fiery trains, the exceeding swiftness with which they pursue their journey among the stars, the rapid disappearance of even the grandest of these seeming chaotic worlds, have all combined to invest these bodies with a power to excite a kind of superstitious terror which even the exact revelations of science cannot wholly dispel. History records the appearance of these phenomena, and in general they were regarded as omens of some terrible scourge to mankind, the precursors of war or pestilence or famine, or at the very least announcing the death of some prince or potentate. Some of the ancients, of course, rose above these superstitious ideas, and the

Roman philosopher Seneca even entertained the opinion that these erratic bodies would some day fall within the domain of human knowledge, that their paths among the stars would eventually be traced, and that they would be found in the end to be permanent members of the solar system. How remarkably this prediction has been verified will appear in the concise sketch we are about to present.

The discovery of the law of universal gravitation was followed by a mathematical demonstration, also accomplished by the great English philosopher, which was the reverse of the problem he had just solved, and may be announced as follows: *Given, the intensity of a fixed central force, decreasing in power as the squares of the distances increase, and the direction and intensity of an impulsive force operating to set in motion a body subject to the central power: Required, the nature and figure of the path described by the revolving body?*

Previous to the resolution of this problem Newton naturally expected to find the curve sought to be an *ellipse.* The sun was the source of a fixed central force which obeyed the above law. The planets were retained in their orbits by this central force. These described ellipses in their revolution around the sun, and it was natural to conclude that the solution of the inverse problem would lead to the elliptic orbit. On completing the solution and reaching the mathematical expression representing the orbit, it was found not to be the usual expression for the ellipse, and after careful examination proved to be the general expression, embracing within its grasp no less than *four* curves, the *circle*, the *ellipse*, the *parabola*, and the *hyperbola.* These curves are allied in a most remarkable manner, having certain properties in common, and having in one sense a common origin. They may all be obtained by cutting the surface of a cone by a plane passing in different directions, as may be seen from the figure below. Let A be the vertex, and C D E

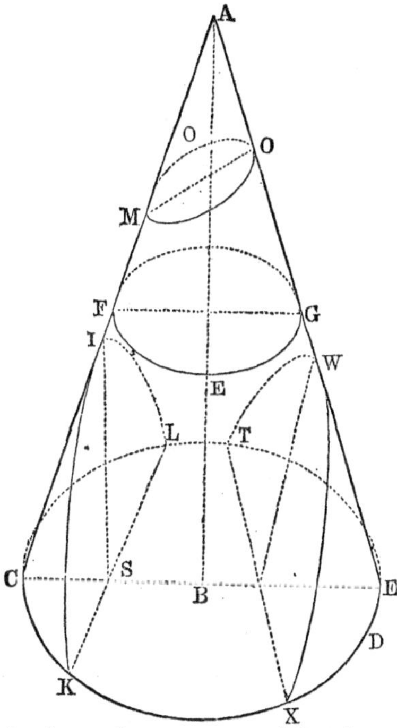

L the circular base of a cone seen obliquely. Any plane passed parallel to the base, or perpendicular to the axis A B, will cut from the surface a circle, as F E G. A plane passed obliquely to the axis will cut from the surface an ellipse, as M O O′. Any plane passed parallel to the side of the cone A C will cut the curve T W X, called a *parabola*, and any plane passed parallel to the axis of the cone A B will cut out from the surface the curve K I L, called an *hyperbola.* These curves are thus all derived from the conic surface by intersecting it with a plane, and are hence called *conic sections.* Now, a little examination will show us that while the circle and ellipse are *re-entering* curves of limited extent, this is not the case with the parabola and hyperbola. If the conic surface were indefinitely extended below the base, it is evident that the cutting plane X W T, being parallel to the side A C, could never cut that particular line, and hence the parabola, departing from the point U, and passing through T and X, would extend indefinitely on the surface of the cone without ever coming together, though the curves would approach each other for ever. Thus the parabola is the limit of all possible ellipses; for it is manifest that as the cutting plane becomes more and more nearly parallel to the side A C, the axis of the ellipse cut out grows longer and longer, and just at the point where parallelism is reached the parabola is formed, and it is only an ellipse with an infinitely elongated axis.

While it is seen that the branches of the parabola approach each other, and may be said to come together at an infinite distance from the vertex at W, this is not the case with the branches of the hyperbola. Departing from the vertex I, and passing the points K and L, the branches of the hyperbola recede from each other for ever, losing by slow degrees their curvature, until at an infinite distance the curves degenerate into straight lines, and thus continue to recede for ever. Such are some of the general characteristics of these remarkable curves. They all, like the ellipse, have a major axis, on each side of which they are symmetrical. They all have at least one *focus*, possessing special properties. They all have a *vertex* lying at the extremity of the major axis and the nearest point of the curve to the focus; and, strange as it may seem, in either one of these curves mathematical analysis demonstrated that a satellite of the sun might revolve under the law of universal gravitation. The elliptic orbits of the planets and the circular orbits of some of the satellites of Jupiter presented examples in the heavens of two of these curves, and it occurred to the sagacious mind of Newton that the hitherto unexplained eccentricity of the cometary revolutions might be accounted for by finding that they revolved around the sun in ellipses of great eccentricity, or possibly in parabolic, or even in hyperbolic orbits. The English astronomer had the opportunity of putting to the test this grand idea by the appearance of a great comet in 1680, which displayed a train of light of wonderful dimensions and seemed to plunge nearly vertically downwards from the pole of the ecliptic, made its perihelion passage with almost incredible velocity, and with a speed always diminishing as it receded from the sun again swept out into the unfathomable depths of space. To this comet Newton first attempted to apply the law of gravitation, and to assign it an orbit among the conic sections. This could be done in the same manner from observation as in the case of a planet. Having obtained as many places of the comet as possible among the fixed stars, it remained to see whether any elliptic orbit or any parabolic orbit could be assigned the comet which would at the same time pass through all these observed places. If this could be done, then it would become possible from this known orbit to predict the places of a comet as of a planet, and in the event of the orbit proving elliptic, then the return of the comet to its perihelion might be computed and announced.

The comet of 1680 was carefully studied by Newton, and its orbit was found to be an extremely elongated ellipse, approaching very nearly to the form of a parabola, but while its physical features and its near approach to the sun made it an object of extraordinary interest, the exceeding velocity with which it swept around the sun rendered it difficult to execute exact observations, and hence this comet was not well adapted to demonstrate the truth of the rigorous application of the law of gravitation to the orbital movements of these eccentric bodies. Another great comet appeared two years later, in 1682, to whose history there attaches a special interest, on account of the fact that it was the first of these bodies shown to have a permanent orbit in connection with the solar system, and the first whose periodic time was sufficiently well computed to render it possible to predict its return. This comet bears the name of the great English astronomer Halley, to whom we are indebted for the computation of the elements of its orbit—a problem, at the time it was executed, far more difficult than any belonging to the whole range of physical astronomy.

The elements of the orbit of a comet are nearly identical with those which fix the magnitude and position of a planetary orbit. To obtain the magnitude of the cometary ellipse we must have two elements, the *length* of the major axis and the *perihelion distance.* To obtain the direction of the longer axis we must have the position of the *perihelion point;* this point, being joined to the sun's center, gives the direction of the major axis. To obtain the position of the plane of the orbit we must have the place of the *ascending* or *descending node*, and also the *inclination* of the plane of the cometary orbit to that of the ecliptic. If, in addition to these elements, we have the *time* of *perihelion passage*, then it becomes possible to follow the comet in its erratic movements with a certainty almost as great as that with which the orderly movements of the planets are pursued.

On the appearance of the great comet of 1682, Halley undertook the laborious and hitherto unaccomplished task of computing rigorously the elements of its orbit, which task he accomplished after incredible labor in the most masterly manner. It then occurred to him to gather up all historic details with reference to the appearance of comets, as well as all astronomical observations, so that by examination and inter-comparison he might learn whether any recorded comet had ever pursued the same track in the heavens which had just been passed over by the comet of 1682. In the course of this historical investigation he found that comets, somewhat resembling in physical appearance, and traversing nearly the same regions of space passed over by his own comet, had appeared in the years 1531 and 1607, and now again in 1682. These epochs are separated by an interval of between seventy-five and seventy-six years, and Halley, after long and laborious computation, announced that in 1759, three quarters of a century from the date of the prediction, this same comet would again return to our system! We can readily sympathize with the feelings of this great astronomer when we find him appealing to posterity to remember, in the event of his prediction being verified, that such an occurrence as the return of a comet was first announced by an Englishman. As the year 1759 approached, the prophetic declaration of Halley excited an unusual interest throughout the astronomical world. To predict the exact point in the heavens to which the telescope must be directed to catch the first faint glimpse of the returning stranger, and to give the date of its perihelion passage, required investigations of so high an order, that in case they had been demanded of Halley, seventy-six years before, the then existing condition of mathematical and physical science would not have furnished the means for their accomplishment. The whole subject of planetary perturbations had by this time been tolerably well developed, and the laborious task of computing the disturbing influence of Jupiter and Saturn was undertaken by Clairault, assisted by La Lande and by a lady, Madam Lepaute, whose name stands in honorable union with the two profound mathematicians. After many months of indefatigable labor the computers announced that for want of time they had been compelled to omit several matters which might make a difference of thirty days, one way or the other, in the return of the comet, but that within these limits this long lost celestial wanderer would pass his perihelion on the 13th April, 1759. The limits of error were justly chosen, for the comet actually returned and passed its perihelion on the 12th of March, just a month ahead of the predicted time.

This successful computation settled for ever the doctrine of the cometary orbits, and demonstrated beyond doubt their subjection to the attractive power of the sun, and that this orb extended its influence into the profound depths of space, to which the comet descended during its journey of seventy-six years. It was further established that Halley's comet was a permanent member of the solar system, performing its orbital revolution around the sun in an exceedingly elongated ellipse, but with a regularity equal to that of the planets. It was further determined that the entire mass of the comet was very inconsiderable, as no account of this mass was made in the computations for perturbation, while the masses of Jupiter and Saturn required to be known with precision. This comet has returned a second time since its discovery by Halley, when its elements were more accurately obtained by many modern astronomers, and perhaps best of all by *Hermann Westphalen*, who predicted its perihelion passage, after an absence of seventy-six years, to within *five days!* This appearance took place at the close of 1835. We shall have occasion to recur to this comet again when we come to speak of their physical constitution.

Westphalen furnishes the following as the actual dimensions of Halley's comet:—

Perihelion distance, ...	55,900,000 miles.
Aphelion "	3,370,300,000 "
Length of the major axis, ...	3,426,200,000 "
Breadth of the orbit, ...	826,900,000 "

It is thus seen that in its journey from the sun this comet crosses the orbits of all the known planets, and passes the boundary of Neptune more than three hundred millions of miles.

Having thus demonstrated the subordination of these extraordinary bodies to the law of universal gravitation and to the received laws of motion, we will proceed to examine their

PHYSICAL CONSTITUTION.—The solid earth we inhabit, the moon, her satellite, the sun and all the planets, are compact masses of matter of differing densities, but of firm, compact materials. The comets, on the contrary, as a class, seem to be vaporous masses, far more unsubstantial than the lightest summer cloud, and in general transparent, or at least translucent, even in their most condensed portions. This is evident from the fact that the minute stars are still visible with undiminished light when seen sometimes through a depth of cometary matter millions of miles in extent. Comets in general consist of a *nucleus* or *head*, the center of force and the most condensed portion of their matter. Around this head there is seen usually a vaporous *envelope* or atmosphere of greater or less extent, sometimes evidently divided into concentric layers of nearly globular form. Many comets, on approaching the sun, undergo extraordinary physical changes in the head or nucleus, which experiences an excessive agitation, flinging out jets or streams of fiery light in a direction towards the sun, which assume many and strange forms, sometimes spreading out into a fan-shaped figure, and rapidly fading in intensity, as they recede from the nucleus. This phenomenon is almost invariably attended or followed with another even more remarkable—the throwing off a train of luminous matter, called the *tail*, in a direction *opposite* the sun, and sometimes extending to a prodigious distance. Thus the tail of the great comet of 1680, already mentioned, according to the computations of Sir Isaac Newton, reached to a distance of more than 140 millions of miles, while only two days were occupied in projecting this inscrutable and mysterious appendage to this enormous distance. The form of the tail is usually that of a hollow paraboloid, the nucleus occupying the focus, and thus the tail, as it recedes from the head, seems to diverge into two streams of light, while the axis or central line is comparatively dark. Sometimes, as in the great comet of 1858, the region immediately behind the nucleus on the axis is *jet black*—the intensity of this blackness growing less and less along the axis until it finally fades out in the general luminosity of the tail.

The nucleus is sometimes tolerably well defined, and presents a planetary disk of greater or less magnitude. It is not intended to assert that there are no comets which are solid bodies, at least in some portion of their central masses. Indeed, if we are to credit the records, some have been seen in the act of crossing the disk of the sun, when they have appeared as round, well defined, circular black spots, exactly like the planets Venus and Mercury, when seen in the same condition on the bright surface of the sun. For the most part, however, we know that these bodies do not present any evidence of solidity. Their heads or nuclei are ill defined when examined by powerful telescopes, and their gaseous condition is demonstrated by the fact that they expand and contract their dimensions with great rapidity, according to circumstances.

This contraction generally takes place as the comet approaches its perihelion passage, which is certainly a very curious fact, and quite contrary to what we would expect, as the excessive heat to which a comet must be subjected in perihelio ought (as would seem) to greatly expand its dimensions. It is doubtless owing to the fact that this enormous heat extends its influence so far that the vaporous mass is expanded and rarified to such a degree as to become absolutely transparent and invisible, and it is only when released from this intense heat by recess from the sun that a condensation takes place, and thus the seeming dimensions of the comet increases. It is difficult to comprehend how some of these bodies, in their nearest approach to the sun, are not absolutely burned up and dissipated for ever. The great comet of 1680, when in perihelio, was only about 147,000 miles distant from the sun's surface; and admitting that the heat of the sun diminishes as the square of the distance increases, Newton computed that the comet was subjected to a heat 2,000 times more intense than that of *red hot iron.* The great comet of 1843 is computed to have approached the sun's surface to within half the above distance, and Sir John Herschel computes that the intensity of the heat then experienced by this comet was 47,000 times greater than the heat of the sun as received at the earth, or more than twenty-eight times greater than the heat concentrated at the focus of a lens of thirty-two inches diameter, which melted agate and rock crystal, and dissipated these refractory solids into an invisible gas!

After passing under the influence of such intense heat, it seems almost impossible that any well defined form should ever be recovered, and yet the comet of 1680 and that of 1843 finally receded from the sun, the nucleus in some mysterious way slowly gathering up its dispersed particles and sweeping away into the depths of space, a well-defined luminous object, not in any sensible degree injured in its form or magnitude by this fiery ordeal.

The *envelopes* of comets and their *tails* are by far the most inscrutable problems of nature. Of these phenomena no satisfactory account has yet been rendered. The envelopes of the comet of 1858 were beautiful in form, with a well defined circular outline, in whose center the nucleus blazed with its fiery light. The diameter of this seemingly globular mass changed from night to night. Its texture varied; sometimes evenly and beautifully shaded and gauze-like in its surface, and sometimes this gauzy surface broken by dark and irregular patches. A second concentric sphere became visible, fainter in its outline than the interior one, and finally a third circle dimly presented its outline, very faint, and only to be seen in powerful telescopes, under favorable circumstances.

The beautiful forms exhibited in these envelopes and retained by them seems to demonstrate the existence of some central repulsive force located in the nucleus, and capable of holding these gaseous particles in equilibrium. What this force may be it is vain to conjecture. If the envelope of the nucleus is a phenomenon surpassing the reach of human thought, what shall we say of the still more mysterious and incomprehensible phenomena presented in the tails of comets?

We have already said that these tails are thrown off in a direction *opposite* the sun as the comet approaches its perihelion passage. As the comet sweeps around the sun with almost inconceivable velocity, the tail retains its direction, just as though its axis were a solid bar of iron, passing through the nucleus to the sun and hanging on the center of the solar orb. This bar, extending out to the furthest extremity of the tail, sometimes 120 millions of miles beyond the nucleus, sweeps round angularly with equal rapidity at every point, so that its rectilinear figure is preserved in this tremendous sweep. In case the tail were composed of ponderable particles, obedient to the laws of gravitation and motion, this would be impossible, for if we consider each particle an independent body, describing an elliptic or parabolic orbit about the sun, the laws of their motion would compel the more distant particles to lag behind the nearer ones in angular velocity.

If no comet ever exhibited any other than this peculiar

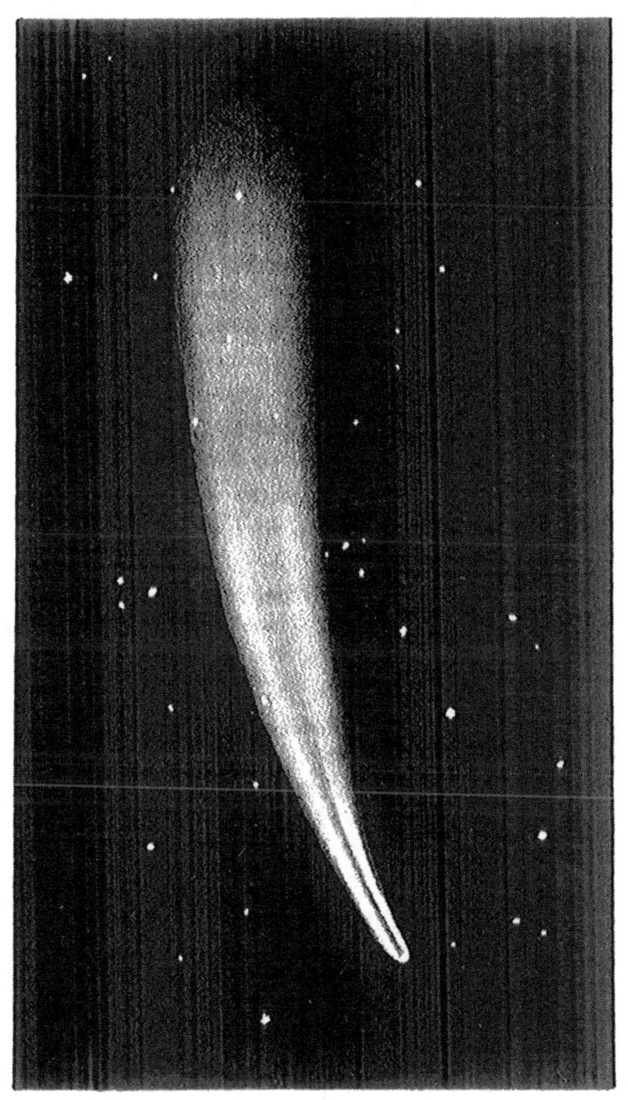

THE GREAT COMET OF 1858.
KNOWN AS "DONATIS COMET"

form of tail, straight and *directed from* the sun, we might frame an hypothesis which could account for the facts; but in some instances there are many tails to the one nucleus, and these not straight, but curved like a cimetar. In other cases there are two tails, the one, as usual, directed *from* the sun, the other pointing towards the source of light. Sometimes the principal tail is straight, and in the direction from the sun, while a lateral ray shooting from the nucleus may form with the axis of the tail an angle of thirty or even sixty degrees.

We have already said that the tail swings around the sun in the perihelion passage, preserving its form and direction, and hence, when the comet is receding from the sun, the tail, in all its vast dimensions, is driven before the head of the comet, preceding the nucleus as it sweeps outward into space.

In some instances corruscations have been noticed to take place in these grand but mysterious appendages, darting with incredible velocity from the very nucleus to the extremity of the tail, and thus flashing backwards and forwards like a magnificent auroral display.

The question arises, What are these luminous displays? Are the tails of comets composed of ponderable matter? If so, do they yield obedience to the known laws of motion and gravitation? Is there any matter in the universe which may ever become luminous, but is imponderable? Can these tails be a mere effect produced on the waves of light emitted by the sun in passing through the mass of cometary matter? These and many other questions equally difficult present themselves in this connection. The re-absorption of the tail into the head would seem to demonstrate that the matter composing the tail was ponderable, while the facts already stated as to the rigid form preserved by the tail in sweeping around the sun positively contradicts this hypothesis.

One thing we know: cometary matter *is* ponderable matter, and obeys the laws of motion and gravitation, is swayed by the sun and by the planets, and in all particulars complies with the laws governing other ponderable matter. This we know, because, as we have seen, it is possible to predict the return of a comet revolving even in so great a period as seventy-five years, and such predictions have been rigorously verified. In case any portion of this ponderable matter were absorbed in the sun, or dissipated by the intense heat which it suffers in the perihelion passage, then would the mass of the comet grow less at each return, and the periodic time would slowly diminish. There is one comet, named after its illustrious discoverer, Encke, whose history for the past thirty years has been followed with high interest, because it is now a fixed truth that at each return its perihelion passage is accelerated by about two and a half hours. It revolves in an elliptical orbit of small dimensions comparatively, and performs its revolution around the sun in a period of only 1,205 days, or about three and a third years. By assuming the existence of a rare *resisting medium*, Professor Encke has succeeded in accounting for the acceleration in the motion of these comets, and this hypothesis has been generally received. In case its truth becomes established it involves remote consequences from which the mind naturally revolts; for if there be a medium capable of destroying any portion of the velocity of Encke's comet, the same resistance must in like manner destroy a part of the orbital velocity of every planet and satellite, and sooner or later each in its turn must by slow degrees approach the sun, and in the end this grand central orb must become the grave of every planet and satellite and comet! Such an hypothesis is combatted, possibly disproved, by the fact that its influence has not yet been discovered on any one of the planets or on any satellite. It may be argued that on these solid substantial bodies it would require ages to produce sensible effect, while on the vaporous ethereal mass of Encke's comet even an almost evanescent medium might produce a sensible effect, even in a single revolution of 1,205 days. May it not be possible to account for the decrease of the periodic time of Encke's comet without having resort to an hypothesis involving the destruction of the entire universe? In case we admit that it loses a portion of its ponderable matter at each perihelion passage, then there must result an effect like the one observed, the comet slowly approaching the sun, to be dissipated entirely, however, before absolutely falling on the surface of the central orb.

However, it is useless to speculate. The facts now in our possession are not sufficient to enable us to render a satisfactory account of the various phenomena in the physical constitution of these bodies which have been enumerated, and we can only hope that the diligence and pertinacity with which this branch of astronomy is now pursued, may before long eventuate in removing from the science this only source of doubt and uncertainty.

In the meanwhile the conclusions reached by Sir John Herschel, from an extended and careful observation of all the phenomena presented by Halley's comet in 1835, 6, have been strengthened by the facts recorded both in Europe and America of the great comet of 1848. All the observations go to demonstrate—

1. That the surface of the nucleus nearest the sun becomes powerfully agitated, and finally bursts forth into luminous jets of gaseous matter.

2. That this matter, with an initial velocity driving it towards the sun, is by some unknown repelling force driven backwards from the sun, and drifted outward from the sun to vast distances forming the tail.

3. That a portion of this vaporized material is not subject to this repulsive force, but remains under the influence of some equally inscrutable central power lodged in the center of the nucleus, and forming the corona or envelope, and assuming forms of great delicacy and beauty.

4. That the force which ejects the tail cannot be gravitation, as it acts with a power and in a direction opposed to this central power.

5. That the power lodged in the nucleus, and by whose energy the particles composing the tail are again reabsorbed into the head, cannot be gravitation, as the minute mass of the comet could not by its gravitating power bring back the particles flung off to such enormous distances.

In this catalogue of inscrutable phenomona we must place the remarkable fact of the splitting up of a comet into two distinct portions. A comet of short period, known as Biela's comet, revolving in about six and three quarter years, was recognized as early as 1826, as a permanent member of the solar system.

This comet, at its appearance in 1832, excited a profound sensation, in consequence of the prediction that it would *cross the earth's path*, thereby creating the greatest alarm among the ignorant lest this crossing might occasion a collision between the comet and the earth.

The prediction was verified, but while the comet was in the act of crossing the earth's track or orbit, the earth was many millions of miles removed from this special point of intersection.

The appearance in 1846 was again rendered memorable by the strange phenomenon already mentioned—the actual severation of the comet into two bodies, distinct and separate, each cometary in its appearance, and each alternately preponderating in apparent magnitude and brilliancy. These two comets possessed all the characteristics which mark these anomalous bodies. Each had its nucleus, its envelope, and its tail. The first indication of a separation occurred as early as the 19th December, 1845. By the middle of January, 1846, the separation was complete, and was well observed in Europe and America. By the beginning of March the interval had increased to a maximum, when it was about one-third as great as the apparent diameter of the moon. From this time the companion comet began to fade, remaining faintly visible up to the 15th March. After this the old comet remained single, and finally disappeared.

Here we have phenomena of the most extraordinary character. What convulsion could have split this nebulous mass into two distinct fragments? What wonderful power could have occasioned the alternations in the intensity of their light? What mysterious bond could have united these severed and separated bodies, and caused them to vibrate about their common center of gravity? Have these bodies been permanently re-united? or will they ever appear as individual and independent objects? These questions it is now impossible to answer.

THE NUMBER OF COMETS far exceeds that of the planets

and their satellites, and, indeed, judging from the list of recorded comets, and taking into account the fact that multitudes of these bodies must escape notice entirely by their remaining above the horizon in the day time, we are forced to the conclusion that they are not to be numbered by hundreds or thousands, but probably by millions! They seem to obey no law as to the inclination of their orbits or the direction of their motions. Some appear to plunge vertically downwards from the very pole of the ecliptic, while others rise upward from below this plane in a direction diametrically opposite. Their planes are inclined under all angles, and their perihelion points are at all distances from the sun. Some revolve in orbits of moderate eccentricity, while others sweep away into space in parabolic or even in hyperbolic orbits, never again to visit our system unless arrested and diverted from their path by some disturbing power. The mighty depths to which some of these bodies penetrate into space, sweeping, as they must, vastly beyond the boundary of the planetary system, would excite a doubt in the mind as to whether there might be *room* enough in space for the undisturbed revolution of these wonderful objects. We shall see hereafter that profound investigations have answered this inquiry and dispelled every doubt as to the grandeur of the scale on which the universe is built.

In the Appendix will be found the elements of the orbits of such comets as are regarded permanent members of the solar system.

We here close our examination of the various classes of attendants on the solar orb. We find this mighty system of revolving worlds composed of bodies which are diverse in their physical constitution, some more dense and solid than the earth on which we dwell, some far more rare and unsubstantial than the atmosphere we breathe—all obedient to the grand controlling power of the central orb, while no one is relieved from the disturbing influence of every other—a vast complicated display of celestial mechanism, whose equilibrium and stability presents the grandest problem for human investigation to be found in the whole universe of matter.

CHAPTER XV.

THE SUN AND PLANETS AS PONDERABLE BODIES.

[GEN]ERAL CIRCUMSTANCES OF THE SYSTEM.—THE SUN.—HIS DIAMETER AND MASS.—GRAVITY AT THE SURFACE.—MERCURY.—HIS MASS AND PERTURBATIONS.—VENUS AS A PONDERABLE BODY.—LONG EQUATION OF VENUS AND THE EARTH.—THE EARTH AND MOON AS HEAVY BODIES.—FIGURE AND MASS OF THE EARTH.—PRECESSION.—ABERRATION.—NUTATION.—MARS.—HIS MASS AND DENSITY.—GRAVITY AT HIS SURFACE.—THE ASTEROIDS.—JUPITER'S SYSTEM.—SATURN.—HIS MOONS AND RINGS AS PONDERABLE BODIES.—URANUS.—NEPTUNE.—STABILITY OF THE WHOLE SYSTEM.

HAVING now completed a rapid survey of the bodies which owe allegiance to the sun, and having reached to a knowledge of those laws which extend their empire over all these revolving planets, we come to the consideration of the modifications which are introduced into the circumstances of motion of each of these worlds, by the fact that it is subjected to the influence of all the others. As under the great law of universal gravitation every particle of matter in the universe attracts every other particle of matter with a force which varies inversely as the square of the distance and directly as the mass, it follows that each planet and comet and satellite of the entire system of the sun are, to a greater or less degree, affected by the attraction of every other.

We have already considered generally the great problem of the "three bodies"—a central, a disturbing, and a disturbed body. The train of reasoning there presented is now to be carried out and extended in succession to the planets and their satellites. Before proceeding to examine the changes wrought in the orbits of the planets and their satellites by the action of all the disturbing forces, we will make a more general examination of the various elements of the planetary orbits, to learn, if possible, whether any of these elements are subjected to changes which are merely periodic in their character, returning after intervals, longer or shorter, to their normal condition, to repeat the same changes in the same order for ever. We desire also to inquire whether any of the elements are subjected to perturbations which always progress in the same direction, and if so, whether these changes in any way involve the destruction of the system as such.

This is undoubtedly the grandest problem ever propounded to the human mind, for it is neither more nor less than an inquiry into the perpetuity of the great scheme of worlds dependent on the sun. It demands a vision which shall penetrate the future ages to predict the mutations and their effects at the end of these ages. It will not be expected that in such a treatise as this we are to enter into an exhaustive discussion of this great subject. We can do little more than announce the results reached by the profound investigations of the great mathematical successors of Newton.

We shall commence, then, by an inquiry touching those elements of an orbit which involve the well-being of a planet, or its fitness to sustain the animal and vegetable life which exists on its surface.

The figure and magnitude of an orbit are determined by the *length* of the *major axis* and by *the eccentricity*, and in case but one planet existed, these are the only elements whose value could in any way affect the physical condition of the planet, so far as its supply of light and heat received from the sun are concerned. In this case the position of the orbit in its own plane (determined by the *place* of the *perihelion point*) and also the position of the plane of the orbit, as referred to any fixed plane, (determined by the *angle of inclination* and *line of nodes*), and also the *epoch* (or place of the planet in its orbit at a given moment of time), all those quantities would not in any degree affect the actual condition of the planet; but as no planet is isolated, and as each is subjected to the influence of every other, it becomes a matter of grave importance to ascertain whether there be any fluctuations in the values of all these elements, whether these fluctuations are confined within any specific limits, and whether, if thus confined, any injurious effect can result to those elements which involve the well-being of any planet; and finally, whether there be any guarantee for the perpetuity of the planetary system in the condition now existent.

In case the planes of the orbits of all the planets were coincident, then the investigations would be confined to the fluctuation in the values of the major axes, eccentricities and perihelia; but from the reasoning already presented in the problem of "the three bodies," we have seen that if we consider the relation of two planets whose orbits are inclined under any angle, in their reciprocal influence, if we assume the plane of the orbit of one of these planets, for example, the earth, as fixed in position, and the plane of the other planet's orbit (as Mars) as inclined to this, under a given angle, it is clearly manifest that the disturbing influence of the earth on Mars will be reversed when Mars passes through the plane of the earth's orbit. Suppose we could place our eye in the prolongation of the line of intersection of the two orbits, then we should see them as two straight lines, inclined to each other, as in the figure below, in which S represents

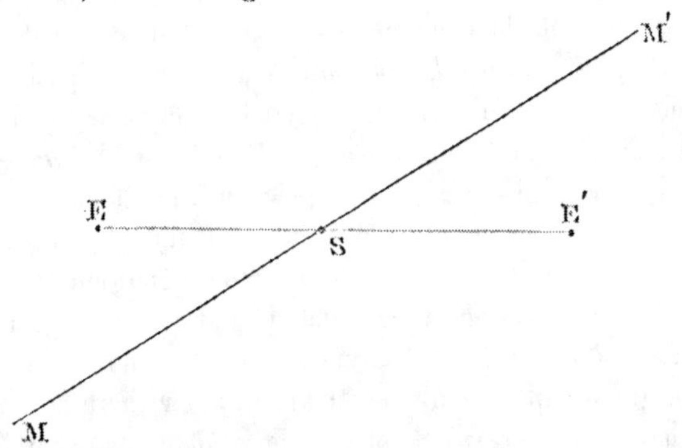

the place of the sun, E E′ places of the earth, and M and M′ places of Mars. Now, when Mars and the earth are on the same side of the sun, or the line of nodes, Mars ascending towards M′, *above* the plane of the earth's orbit, the force exerted by the earth on the ascending planet tends to draw it downward to the ecliptic, and hence it will not quite reach the elevation M′, and thus all this while the plane of the orbit of Mars will be forming a less and less angle with the ecliptic. The moment, however, the planet reaches its highest elevation and commences to descend towards the ecliptic to pass its descending node, then the earth, remaining stationary, will pull the planet towards its own plane, and hence the descent will be made steeper, and the angle of inclination of the orbit of Mars will, while the planet thus descends, be always *increasing*. Following Mars below the plane of the ecliptic, the earth remaining as before, we see that here the tendency is to pull Mars up to the earth's orbit, and hence it will not quite reach the point M, or the inclination in this descent below the ecliptic will *diminish.* From this point, as Mars begins to ascend, the earth's attractive energy will cause it to ascend more rapidly, and will make it pass its ascending node earlier than if undisturbed, and as it comes up faster, it must ascend a steeper grade, or the *inclination* will increase. Thus we see that Mars, in ascending or descending to pass either node, will both ascend and descend by a steeper grade because of the earth's attraction, while in passing from either node to the highest and lowest points of its orbit the same force will operate to make the planet reach points less remote from the ecliptic than if undisturbed, and hence to ascend and descend with a smaller angle of inclination. Now, a careful inspection will show that the effects produced by the earth on Mars, while situated at E′, will be greater in the half of the orbit of Mars which lies above the ecliptic, and at the end of one revolution an exact compensation may not be effected, so that the *increase* of the angle of inclination may not be exactly equal to the *decrease.* But as the earth is revolving, a time will come when this body will occupy the point E, and then the most powerful effect will be produced when Mars is *below* the ecliptic, and in case the orbits are circular, an exact symmetry existing, at the end of a certain cycle the inclinations will be exactly restored.

The fact that the orbits of the planets are elliptical in figure cannot in any way lessen the force of the reasoning we have employed; it can only postpone to a more remote period the final restoration of the inclinations of the planetary orbits. Under the powerful and masterly analysis of Lagrange this subject was completely exhausted, and a result reached which in the following proposition guarantees the stability of the inclinations through all ages:—

"If the mass or weight of every planet be multiplied by the square root of its major axis, and this product be multiplied by the tangent of the angle of inclination of the plane of the planetary orbit to a fixed plane, and these products be added together, their sum will be constantly the same."

Now, we will show hereafter that the major axes remain nearly invariable, the masses of the planets are absolutely so, and hence the third factor of the product, *the tangent of the inclination*, can only vary within narrow limits, returning at the end of a vast cycle to the primitive value.

We shall see hereafter how important the stability of the inclination of the earth's orbit is to the well-being of the living and sentient beings now on the earth's surface.

We proceed to examine the changes of the *lines of nodes* due to perturbation. These changes are allied to those of inclination, and are, indeed, a necessary consequence of these changes, as may readily be shown.

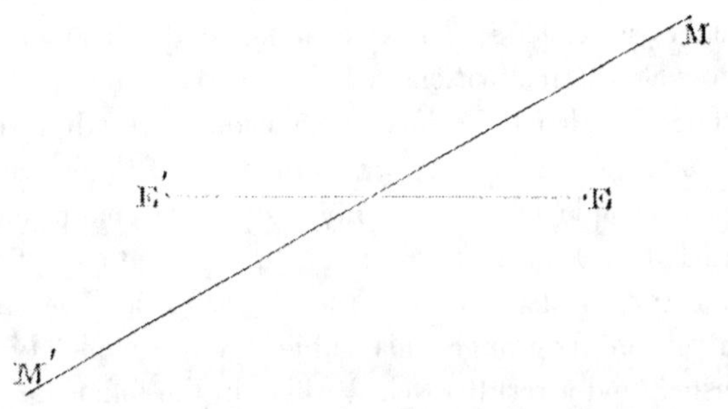

For this purpose we return to the figure already employed, using the same planets, Mars and the earth, regarding the movements of Mars to be disturbed by the earth's attraction.

We have already seen that in case Mars be at M, the earth being at E, the planet, in descending its orbit to the line of nodes seen as in S, (the eye of the spectator being in the prolongation of the line of nodes) during the entire descent the planet will be drawn down to the plane of the ecliptic E E′ on a steeper grade than the normal one M S, and hence the planet will pass through the ecliptic earlier than if undisturbed, or at a point which will be seen somewhere between S and E. Thus during this descent the node will go backwards to meet the planet, or will *retrograde.* Passing below the ecliptic, the planet continuing to descend, will, as we have seen, be prevented by the disturbing body from reaching a point so low as M′, and hence if its path for a moment were anywhere produced backwards, this line would meet the ecliptic at some point always approaching E, or here again the line of nodes retrogrades. The same reasoning will show that with the above configuration the retrogradation of the line of nodes must continue with unequal velocity during the entire revolution of the planet. In other configurations there is sometimes an advance of the node, but in the long run it is easily demonstrated that the nodes of all the planetary orbits on any fixed plane will retrograde and perform entire revolutions in periods of greater or less duration.

This perpetual recess of the lines of nodes in one direction does not in any way affect the physical condition of a planet, but serves an admirable and necessary purpose in securing final stability in the planetary system by presenting the disturbed orbits to the disturbing bodies under all possible configurations.

We shall not attempt to exhibit in full the reasoning by which the variations of the remaining elements are shown to be periodic, when periodicity is essential to stability, or progressive when progression does not involve destruction, but from a single figure deduce, if possible, the great principles involved in this wonderful problem.

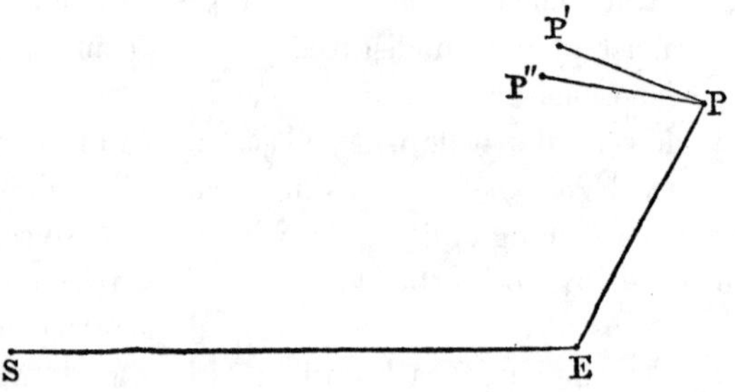

Let S represent the sun, E the earth, and P any planet disturbed by the earth, and let us suppose that undisturbed in any small portion of time it would reach P′, but subjected to the influence of the earth's attraction it reaches P″ in the same time. The question is, in what way does this change affect the elements of P's orbit?

We have already seen the effect on the *inclination* and *line of nodes*. These elements do not affect the magnitude of the orbit nor the position of that orbit in its own plane. The magnitude and position depend on the length of the major axis, eccentricity and perihelion point. Let us examine these in order, commencing with the length of the *major axis*.

We suppose the planet P to be moving, when undisturbed, with its normal elliptic velocity, and of course on reaching P′, the longer axis of its orbit, and in fact all the elements remain unchanged in value; but being disturbed, so as to be prevented from reaching P′, and being compelled to reach P″, will this compulsion merely change the position of the planetary orbit, or will it increase or decrease the length of the major axis?

Kepler's third law tells us that the squares of the periodic times are proportional to the cubes of the major axes, and from this relation it is manifest that any change in the elliptic velocity of a planet must change the period of its revolution, and this involves a change by necessity in the major axis of the orbit.

The question of change in the major axis, then, resolves itself into an inquiry as to whether the disturbance has produced any change *in the elliptic velocity* of the disturbed planet.

At the first glance it may seem impossible to drag a planet from its normal elliptic path without affecting its velocity. This, however, is not the fact. If a body be moving in a straight line and a force be applied to it perpendicular to the direction of its motion, this force will not in any degree affect the velocity, but only the direction of the moving body. Thus a ball fired from a rifle on the deck of a fixed or moving boat, with the same initial force, will reach the opposite shore in the same time, but its direction of absolute motion is changed if fired from a moving boat, from what it would be if shot from one at rest. So a flying planet may be subjected to the action of a force always perpendicular to the direction of its motion, which force may push it from its normal path, but cannot affect its elliptical velocity.

Such a force, then, can have no influence on the length of the major axis, or on the periodic time of the revolving body.

Now, every force is capable of being changed into *three* other forces whose combined action will produce the same effect as the primitive force, as in the figure.

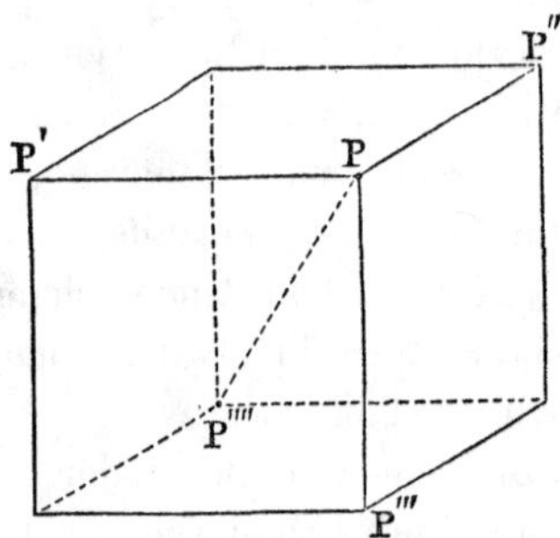

Let P P'''' represent the direction and intensity of any force; on this line as a diagonal construct the solid figure a parallelopiped. Then the sides P P', P P'' and P P''' will represent the direction and intensity of three forces, which would produce the same effect as the force P P''''.

Precisely in this way the disturbing force exerted by the earth on the planet P can be converted into three other forces, whose combined effect shall be identical with the original force. Two of these forces shall lie in the plane of the planet's motion, the one tangent to the orbit, or in the exact direction of the planet's motion, the second perpendicular to the direction of motion, or normal to the orbit, and the third perpendicular to the plane of the orbit of the planet.

Now, from what we have said, it is clear that but one

of these new or substituted forces can in any way affect the velocity of the planet, and that is the force tangent to the orbit, or coincident with the direction in which it is moving. The normal component (as it is called) pushes the planet from its old orbit and the perpendicular component pushes the planet above or below its own plane of motion, but neither of these affect the velocity of the moving planet, and neither of them can in any degree affect the length of the major axis.

The perpendicular component has already been considered in its effects, for it is this force which changes the *inclination* and gives motion to the line of nodes. We may, therefore, in our future examinations leave this force out of consideration, or, which comes to the same thing, consider the planes of the orbits of the disturbing and the disturbed planet as the same.

Let us, then, represent by the two circles the orbits of the planets in question, S being the place of the sun,

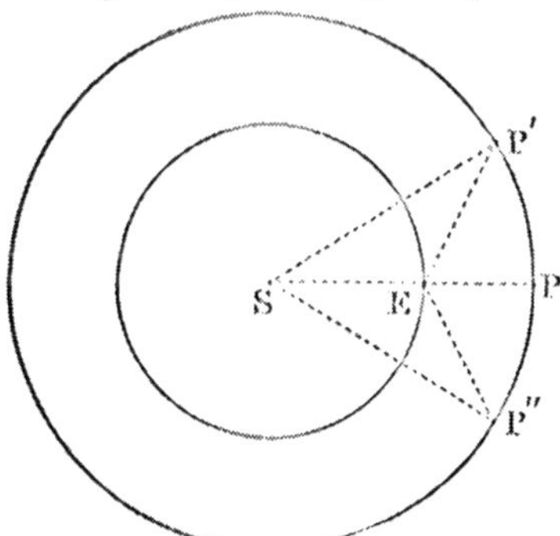

E the earth, and P the planet when in conjunction. In this configuration the entire disturbing power of E is exerted along the line E P, or perpendicular to the direction of the planet's motion, or *normal* to its orbit, and in

this position the tangential force being nothing, the major axis is undisturbed by E. As P moves towards P′ the direction of the force exerted by E ceases to be normal to P's orbit, and may be replaced by a normal and a tangential force. The tangential force from P towards P′ is manifestly in opposition to the motion of P in its orbit, and therefore retards its motion, and thus decreases its major axis; but there is a point P″ symmetrically placed with reference to P, where the tangential force is in the opposite direction, and in an equal degree becomes an accelerating force, and whatever the major axis might lose in length from the disturbing power at P′, it would gain from the same power when it comes to occupy the point P″. So that if E should remain fixed during an entire revolution of P, a compensation would be effected, and the velocity of the planet on reaching its point of departure would be identical with that with which it started, and hence the major axis, though it would have lost and gained, would in the end be restored to its primitive value.

If the orbits were elliptical and their major axes were coincident the same reasoning from symmetry would still hold good, and demonstrate the restoration of the major axis, and as action and reaction are always equal, it is manifest that by fixing P and causing E to revolve, the changes wrought by E on P would now be wrought by P on E, only in the reverse order—that is, wherever P was accelerated by E, E will be retarded by P, and *vice versa*.

Admitting the major axes to be inclined to each other destroys the symmetry of the figure, and an exact restoration is not effected in one revolution; but as the perihelia of the planetary orbits are all in motion, the time will come when a coincidence of the major axes will be effected, and if there be a certain amount of outstanding uncompensated velocity, when the coincidence takes place, the action will be reversed, and at the end of one grand revolution of the major axes from coincidence to coincidence the restoration will be completed, and the axes will return to their primitive value.

Here we are compelled to leave the problem, and simply state the result which a complete solution has effected. We are again indebted to Lagrange for the resolution of this most important of all the problems involving the stability of the solar system, who presents the final result as follows :—

"If the mass of each planet be multiplied by the square root of the major axis of its orbit, and this product by the square of the tangent of the inclination of the orbit to a fixed plane, and all these products be added together, their sum will be constantly the same, no matter what variations exist in the system."

The mass or weight of each planet is invariable, while the loss or gain in the values of the major axes is always counterpoised by the gain or loss in the inclination of the orbits, and thus in the long run, in cycles of vast periods, a complete restoration of the major axes is fully accomplished, and the system in this particular returns to its normal condition.

We have thus far considered the effect of two out of the three forces into which a disturbing force may be decomposed. The normal component remains to be examined. This acts in a direction normal to the curve described by the planet, or perpendicular to the tangent to the orbit at the point occupied by the planet. We shall not enter into any extended examination of this subject, and will only say that this component of the disturbing force gives rise to a movement in the perihelion points of the planetary orbits, sometimes advancing these points, sometimes giving them a retrograde motion, and in some instances producing oscillations.

These effects are necessarily mixed up and combined with those produced by the action of the tangential force, for, as we have seen, the effect of this force goes to increase or decrease the value of the major axis; but no increase or decrease of the major axis can take place without a corresponding change in the eccentricity, so that these changes thus modified, the one by the other, finally become exceedingly complex, and can only be traced and computed by the application of the highest powers of analytic reasoning.

The complexity is further increased from the fact that in the consideration of the entire problem of perturbations the varying distances of the disturbing and disturbed bodies must be rigorously taken into account, and may modify and even reverse the effects due simply to direction. With difficulties so extraordinary and diversified, with complications and complexities mutually extending to each other, involving movements so slow as to require ages for their completion, it is a matter of amazement that the human mind has achieved complete success in the resolution of this grand problem, and can with confidence pronounce the changes to fall within narrow and inocuous limits, while in the end, after a cycle of incalculable millions of years, the entire system of planets and satellites shall return once more to their primitive condition, to start again on their endless cycles of configuration and change.

There remains one more source whence arises an accumulation of disturbance, progressive in the same direction through definite cycles of greater or lesser duration. I mean the effects due to a *near commensurability* of the periods of revolution of the disturbed and disturbing planets. The nearer the approach to commensurability the longer will be the duration of the resulting inequality.

We shall have occasion to resume this subject in our examination of the circumstances of disturbance belonging to each individual planet, which we shall now proceed to examine briefly, commencing at the sun, and proceeding outwards.

THE SUN CONSIDERED AS A GRAVITATING BODY.—We shall now return to the great center of the planetary worlds with a full knowledge of the laws of motion and gravitation, and provided with the instrumental means of securing those delicate measures whereby the solar orb may be determined in *distance, volume* and *weight.*

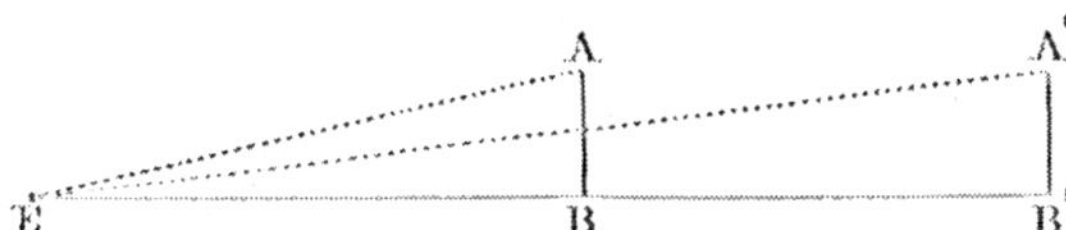

We have already explained how these quantities may be obtained, and we now present the results of exact measures and accurate computation. The sun's mean distance from the earth may be taken at ninety-five millions of miles. By exact measures the mean diameter of the sun subtends an angle equal to 32′.01″.8, and an angle of this value indicates a real diameter in the sun of 883,000 miles, as may be seen from the figure above, in which it is evident that a line A B subtends at A′ B′, a much smaller angle than when located at A B, nearer the vertex. If, then, we have a given angle A′ E B′, and a given distance E B′, from the vertex, if we erect the line B′ A′ it is evident the length of this line will be determined by the value of the angle B′ E A′ and the length of the line E B′; so the sun's distance and angular diameter determines his real diameter. This diameter of the sun (883,000 miles), in terms of the diameter of the earth, amounts to 111.454, and as the volumes of two globes are in the proportion of the *cubes* of their diameters, we shall have the volume of the sun to the volume of the earth as $(111.454)^3$ is to $(1)^3$, or as 1,384,472 to 1, or it would require no less than one million three hundred and eighty-four thousand four hundred and seventy-two globes as large as the earth to fill the vast interior of a hollow sphere as large as the sun. By this we do not mean to assert that the sun weighs as much as 1,384,472 earths. This is not the fact. We have seen already the process by which the relative weights of these globes may be reached, and we have found that the force exerted on the earth by the sun, enfeebled by the distance at which it acts, and thus reduced to the 160,000th part of its actual value, at a distance equal to that of the moon, still exceeds in a more than twofold ratio the force exerted by the earth on the moon; and when the exact ratio is applied we find the weight of the sun to be equal to 354,936 earths, and this is what we call the *mass of the sun.*

If we divide the mass by the volume we obtain the specific gravity of the sun, in terms of that of the earth, equal to $\frac{354,936}{1,384,472}=0.2543$—that is, the average weight of one cubic foot of the sun is only one-fourth as great as the average weight of one cubic foot of the earth.

It is the *mass* of the sun and not its *volume* which determines the amount of force which this great central globe exerts. Its weight is such as vastly to exceed that of any one of the planets, and indeed it rises so superior to the combined masses of all the planets that the center of gravity of the system falls even within the surface of the sun. This may be shown by an examination of the weights and respective distances of the planets. We will explain the reasoning. If the sun and earth were equal in weight, then the center of gravity would lie in the middle of the line joining their centers; but the sun is equal in weight to 354,936 earths, and hence, dividing the distance between the centers (ninety-five millions of miles) into 354,936 equal parts, the center of gravity of the sun and earth will fall on the first point of division nearest the sun's center, that is at a distance of about 267 miles; but from the center of the sun to his surface is a distance of 440,000 miles, and thus the center of gravity of the sun and earth falls far within the limits of the solar surface.

The energy exerted by the sun on any one of his satellites is in a constant state of fluctuation, growing out of the variation in the distance of the planet. The sun's force decreases as the square of the planet's distance increases. If, then, we take the earth's distance as unity, and call the force exerted on the earth by the sun *one*, the force exerted on a planet twice as remote from the sun as the earth, would be but *one-fourth*, at three times the distance *one-ninth*, at ten times the distance it would be but *one hundredth* part of that exerted on the earth. This law of gravitation should be well understood, as we shall have occasion to make frequent applications in our future examinations.

POWER OF GRAVITATION ON THE SOLAR SURFACE.—If it were possible to transport a body weighing at the earth's equator *one pound* to the equator of the sun, as the weight of the body is due to the power of the earth's attraction, and as the sun is heavier than the earth in the high ratio of 354,936 to 1, we might suppose that the pound weight on the earth removed to the sun would be increased in the same ratio. This would be true in case the sun's diameter were precisely equal to that of the earth. This, however, is not the case. The radius of the sun is 111.454 times that of the earth, and this distance will reduce the attractive power of the sun in the ratio of $(111.545)^2$ to $(1)^2$, or as 12,342.27 to 1. If, therefore, we reduce 354,936 in the above ratio, or, in other language, divide it by 12,342.27, we obtain for a quotient 28, showing that a body weighing one pound at the earth's equator would weigh 28 pounds at the sun's surface. This would be slightly reduced from the uplifting action of the centrifugal force due to the velocity of rotation of the sun on its axis. This diminution may be readily computed. We shall see hereafter that the centrifugal force at the earth's equator is equal to $\frac{1}{289}$ of the force of gravity. Now, if the sun rotated in the same time as the earth, and their diameters were equal, the centrifugal force on the equators of the two orbs would be equal. But the sun's radius is about 111 times that of the earth, and if the period or rotation were the same the centrifugal force at the sun's equator would be greater than that at the earth's, in the ratio of $(111)^2$ to 1, or more exactly in the ratio of 12,342.27 to 1. But the sun rotates on its axis much slower than the earth, requiring more than 25 days for one revolution. This will reduce the above in the ratio of 1 to $(25)^2$, or 1 to 625; so that we shall have the earth's equatorial centrifugal force $\frac{1}{289} \times 12,342.27 \div 625 = \frac{12,342.27}{180,625} = 0.07$ nearly for the sun's equatorial centrifugal force. Hence the weight before obtained, 28 pounds, must be reduced seven hundredths of its whole value, and we thus obtain $28 - 0.196 = 27.804$ pounds as the true weight of one pound transported from the earth's equator to that of the sun.

These principles enable us to compute readily the gravitating force exerted by the sun at any given distance; and, as we shall see hereafter, this mighty central orb is pre-eminently the controlling body in the scheme of revolving worlds, which move about him as their center, in obedience to the laws of motion and gravitation.

We close what we have to say of the sun by stating that a heavy body weighing, as it does, 27.8 times as much at the solar as at the terrestrial equator, if free to fall, will pass over in one second a space equal to $27.8 \times 16.1 = 247.58$ feet.

Perturbations of Mercury.—In our discussion of the planets already given we were only prepared to present the discoveries of formal astronomy. These involved the elements of the elliptic orbits and the observed circumstances of the planetary movements. We are now prepared to understand how the system of solar satellites constitutes a grand assemblage of worlds in motion and yet in equilibrio, so that, although there be fluctuations to and fro, which are really perpetual, in the end the system is stable and in exact dynamical counterpoise. The great law of universal gravitation being known, as also the laws of motion, it becomes possible to determine the exact conditions of this mighty systematic equilibrium, and in a strict sense to *weigh* each of the worlds belonging to the system. Indeed, this weight or mass of the planets must be first ascertained before it becomes possible to compute the influence exerted by one body on another, even when their actual distances are known. The distances being the same, two bodies attract a third by a force which is in direct proportion to their masses. Hence, if our moon could be conveyed successively to each of the planets and be located at the same distance from each it now is from the earth, the periods of revolution of our satellite round any one of these worlds would show us whether that world weighed more or less than ours. Thus in case the period of revolution of the moon around a planet should be one-half its present period, then that planet, holding, as it does, the moon with double the velocity at the earth, it must be double the weight of the earth, and so for any other period.

It is in this way that we are enabled very exactly to weigh the planets which are surrounded by satellites, as we have already seen; but those planets which have no satellite, such as Mercury, Venus and Mars, can only be weighed by the effect they produce on other bodies of the system, and especially on those vaporous masses the comets, which occasionally come sufficiently near these bodies to be subjected to very powerful perturbations.

The mass of Mercury is, of course, subject to some uncertainty, but as now determined, in case the sun were divided into *one thousand millions of equal parts*, it would require 2,055 of these parts to be placed in one scale of a balance to counterpoise Mercury in the opposite scale.

Knowing the mass of a planet and its volume, we can easily deduce its specific gravity or density. For example, the volume of Mercury is equal to 0.595, the earth being unity; but the mass of the earth in the same parts of the sun just employed, as we shall see, is 28,173. Hence, if these planets were equally dense, their volumes would be to each other as 28,173 to 2.055, or nearly as 13.7 to 1, or as 1 to 0.072; but Mercury's volume is but 0.595, the earth being taken as unity, and hence Mercury must be denser than the earth in the ratio of 0.072 to 0.595, or as 1.2 to 1 nearly.

Thus are we made acquainted with the very structure or material of the planets by the process of weighing them, revealed by the laws of motion and gravitation, and that these results cannot be much in error is manifest from the fact that the transit of Mercury across the sun's disk, which occurred on the 8th May, 1845, and observed at the Cincinnati Observatory, the *computed* and *observed* contact of the planet with the sun's limb differed by only *sixteen seconds of time!*

This prediction also verifies the values of the secular inequalities, or slow changes in the element of Mercury's orbit, due to the planetary perturbations, which were fixed for the beginning of the present century as follows:—The perihelion makes an absolute advance each year of 5″.8; the node recedes annually 7″.8. The eccentricity, in terms of the semi-major axis, was 0.210.551.494, and its decrease in one hundred years amounts to 0.000.003.866, in terms of the same unit. Let us admit these changes to be progressive at the same rate, and then convert them into intelligible terms, and examine the results. The perihelion point advancing at the rate of 5″.8 a year, will require to pass over

360°, or 1,296,000 seconds, $\frac{1,296,000}{5.8}=223,449$ years. Such is the vast period required for one revolution of the perihelion point. In like manner we may see that the node requires a period exceeding one hundred thousand years for its revolution.

The eccentricity is slowly wearing away, the orbit becoming more nearly circular, and if this change progresses uniformly, it will require no less than five million four hundred and forty-six thousand two hundred years to reduce the figure of Mercury's orbit to that of an exact circle!

The present eccentricity of the orbit of Mercury is such that the aphelion distance exceeds the perihelion distance by more than fifteen millions of miles. The energy exerted by the sun on the planet at perihelion, as compared with that exerted at aphelion, may be readily computed thus: Mercury's greatest distance from the sun amounts to about forty-four millions of miles. This is about ten times the solar radius, and at this distance the sun's power will be reduced to the one hundredth part of what it is at the sun's surface, but the planet when nearest the sun is distant about twenty-nine millions of miles, that is, less than seven times the solar radius, and hence the power of gravitation is reduced to the one forty-ninth part of what it is at the sun's surface, or, what comes to the same thing, Mercury at aphelion is attracted with a force only *one-half* as great as that by which it is affected when nearest the sun.

If a person were transported to the equator of Mercury his weight would be greatly reduced from that found on the earth. The mass of Mercury, in terms of that of the earth as unity, is but 0.729, and if Mercury's diameter were equal to that of the earth, then one pound on the earth would weigh 0.729 lbs. when removed to Mercury; but as the radius of Mercury is only 1,544 miles, or twenty-six hundredths of the earth's radius, this will increase the weight in the ratio of $(2.6)^2$ to $(1)^2$, or as 6.76 to 1. Hence, by multiplying 0.729 by 6.76, we have 0.493 lbs. as the weight of a terrestrial pound removed to Mercury, that is, the power of gravitation on the surface of this planet is about one-half of what it is on the earth.

All the planets exterior to the orbit of Mercury exert an amount of power on this nearest planet to the sun which varies directly as the mass, and inversely as the square of the distance of the disturbing body. Let us suppose the earth and Venus to be in conjunction with Mercury, and that these planets are at their mean distances from the sun, and let us compute in this configuration the relative power of the sun, of Venus, and of the Earth, over Mercury.

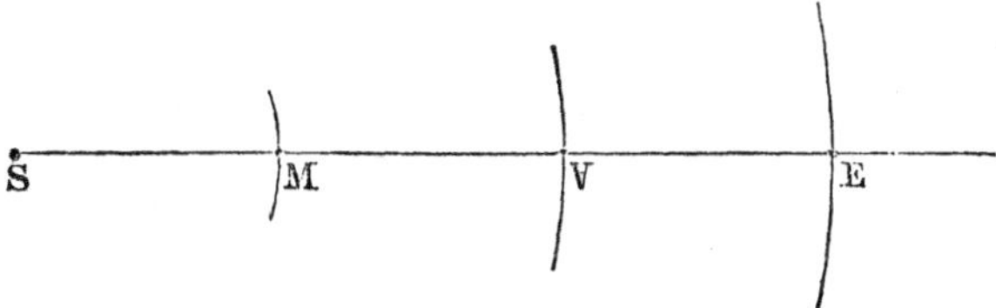

In the figure let S represent the sun, M Mercury, V Venus, and E the Earth. Taking the distance S E to be 1, S M will be 0.387, and S V will be 0.723. Hence, M V will be equal to S V−S M=0.723−0.387=0.336, and V E will be equal to S E−S V=1.000−0.723−0.277. As the mass of Venus is but the 390,000th part of the sun's mass, her effect on Mercury at equal distances would be but one part in three hundred and ninety thousand of the sun's power. The fact that Venus is nearer Mercury than the sun, in the ratio of V M to S M, or of 0.336 to 0.387, will increase her relative power in the ratio of the squares of these quantities inversely, that is, as $(0.336)^2$ to $(0.387)^2$, or as 0.113 to 0.147, or as 1 to 1.3—that is, we must multiply $\frac{1}{390,000}$ by 1.3 to obtain the effect of Venus on Mercury, as compared with that of the sun; in other language, if the sun's power over Mercury be divided into 390,000 equal parts, the power of Venus over the same planet will amount to just one and one-third of these parts.

Let us now compute the attraction of the earth as compared with that of the sun. As the earth weighs 1, while the sun weighs 354,936, at equal distances the powers of the earth and sun would be as 1 to 354,936; but the distance S M is 0.387, while the distance M E is $1.000 - 0.387 = 0.613$. As the sun is the nearer, his power will be increased in the ratio of the square of distance inversely, or as $(0.613)^2$ to $(0.387)^2$, or as 0.376 to 0.147, or as 2.56 to 1—that is, the earth's power, which on account of its mass is but one part in 354,936 of that of the sun at equal distances, must further be reduced on account of its distance to a fraction of this quantity, represented by $\frac{1}{2.56}$, that is, in case we divide the sun's power of attraction upon Mercury into 787,340 parts, the attractive power of the earth will be represented by one of these parts.

We have seen above that the power of Venus over Mercury is equal to $\frac{1.3}{390.000}$, the power of the sun being 1. The power of the earth is $\frac{1}{787.340}$, or not quite one-half of the former quantity. Hence the disturbing influence of Venus is evidently the predominating one in the case of Mercury.

It may be well to extend our investigation a little

further and examine the influence of the massive planet Jupiter on Mercury, to see whether Venus still predominates in its power over that of the heaviest planet of the system. The distance of Jupiter from the sun is 5.2, that of the earth being 1. The distance of the earth from Mercury is 0.613. The distance of Jupiter from Mercury is $5.200-0.387=4.813$. In case the earth and Jupiter were equal in mass, then the power of Jupiter over Mercury would be to the power of the earth as $(0.613)^2$ to $(4.813)^2$, or as 0.376 to 23.164, or as 1 to 61.6—that is, Jupiter's effect is reduced to the fraction $\frac{1}{61.6}$ of what it would be at a distance from Mercury equal to that of the earth; but this is supposing the earth and Jupiter to be equal in mass, whereas Jupiter really requires 338 earths to counterpoise his weight. We must, therefore, increase the fraction $\frac{1}{61.6}$ 338 times, and we have $\frac{338}{61.6}=5.5$ about. Hence, Jupiter exerts a power over Mercury when in conjunction 5.5 times as great as that exerted by the earth, or two and a half times greater than the attraction of Venus.

These computations have been made to show how minute a portion of the sun's power is that exerted by any planet to disturb the motions of another planet, and also to show that we cannot neglect any disturbing body because of the great distance at which it may be placed.

It will be readily seen that when Mercury is in opposition with respect to Venus, her power is greatly reduced on account of the increased distance by which the planets will be then separated. Indeed, the attractive force computed at conjunction will be reduced to about one-tenth at opposition.

This is not true, however, of the attractive power of Jupiter. The distance 4.813 in conjunction will only be increased by the diameter of Mercury's orbit, or by 2(0.387)=0.774 when in opposition, and the distances will stand 4.813 and 6.587. Jupiter's power at the increased distance will be reduced only in the ratio of the square of 4.813 to that of 6.587, or about as 1 to 2.

As an exercise the student should compute the energy exerted by the other planets over Mercury, and thus obtain a familiarity with the application of the law of gravitation to the problems of nature.

While we are writing the intelligence has reached this country that a new planet has actually been discovered, revolving in an orbit between Mercury and the sun. M. Le Verrier some time since announced that there were perturbations in the elements of the orbit of Mercury not explained by any of the known causes, and hence he drew the conclusion that possibly a ring of very small planets were revolving within the limits of Mercury's orbit. One of these minute planets is said to have been actually seen more than once by an amateur astronomer, whose name is M. Lescarbault. This planet is said to complete its revolution in about three weeks, and hence its distance from the sun must be about fourteen or fifteen millions of miles.

VENUS CONSIDERED AS A PONDERABLE BODY.—The angle subtended by this planet at its mean distance from the earth amounts to 17″.55, showing an actual diameter nearly equal to that of the earth. The weight or mass of Venus is not so well determined as that of the planets attended by satellites, yet we have reason to believe that the approximate value does not differ by any very considerable amount from the true one. As now determined by the best authorities, Venus weighs 0.900, the weight of the earth being assumed as 1.000. If an inhabitant of the earth were transported to Venus, his weight would be reduced in the ratio of 1 to 0.94, and a heavy body, free to fall, would pass over 15.1 feet in the first second of time. Here we might repeat the reasoning already employed in the case of the sun to reach these results, but as we shall have occasion hereafter to apply the reasoning to the cases of the larger planets, we shall merely refer to the demonstration already made.

All the elements of the orbit of Venus are in a state of constant fluctuation. The exact condition of these elements will be given hereafter, as well as the measured amount of the changes. We find in Venus the first example of a remarkable perturbation arising from a cause already adverted to, viz.: an approximate commensurability between the periods of revolution of Venus and the earth. The planet performs her revolution around the sun in 224.700 days, while the earth occupies 365.25 days in accomplishing her revolution. If we multiply 224.7 by 13 we obtain 2,921.10; multiply 365.256 by 8, and we have 2,922.048. Thus we perceive that in case Venus and the earth are in conjunction on any given day, at the end of 2,921 days they will be nearly in conjunction again at the same points of their orbits. Whatever perturbation the one planet produces on the other will be again repeated on the return of the same identical configuration. But we have already seen that the synodical revolution of Venus is accomplished in 583.9 days. This quantity, multiplied by 5, produces 2,919.6—that is, during the time involved in the long cycle of 2,921 days there have occurred five conjunctions of the earth and Venus distributed equally around the orbits of the planets. If we examine the figure below, and suppose S, V and E to represent the places of the sun, of Venus and the earth, at the commencement of a great cycle of 2,921 days, at the end of one synodic revolution of Venus V′ and E′ will be the places of the two planets. At the end of the

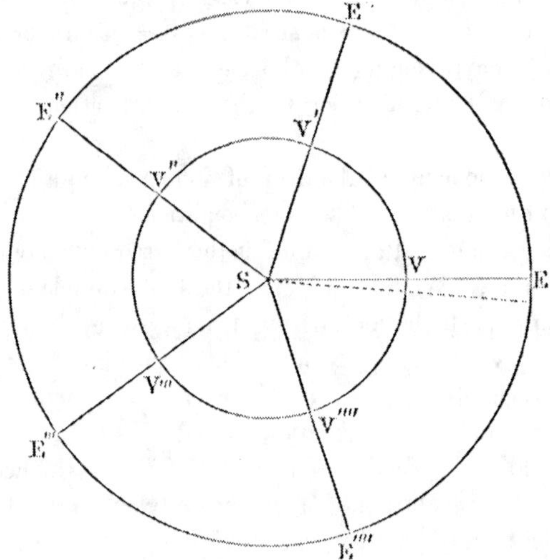

second synodic revolution the planets will be in V″ and and E″, and thus they will pass round the orbits, making their conjunctions at intervals of 583.9 days, and separated by arcs equal to one-fifth part of 360°. Let us now carefully examine the reciprocal influence of V and E. Starting from the places V and E in the figure, Venus will take the lead, and will tend to drag forward the earth, while the earth will pull back the planet, and as the planets sweep around the sun, Venus overtakes the earth at V⁗ E⁗, and as the earth is now in advance, it will accelerate Venus, and will in turn be retarded. Thus a partial compensation is effected, and the motions of the planets return nearly to what they were at the start. This same process is repeated at every conjunction; and in case the planets fall exactly on the right line S V E, at the end of five of these conjunctions a complete restoration would be effected. But this is not exactly true. The periods are not precisely equal, and the fifth conjunction does not fall on S V E, but on a dotted line a little behind the position S V E. There will, therefore, remain a very small amount of outstanding perturbation at the close of one great cycle, which will go on accumulating so long as the dotted line falls in the same half of the earth's orbit. But the difference between 2,921.160 days and 2,922.048 is 0.852, and by this fraction of one day is the earth later than Venus in reaching the point of departure. Hence, the conjunction of the planets must have taken place on a line behind that of the former conjunction, whose position may be readily computed. The daily motion of Venus is $1^{\circ}.612$, while that of the earth is $0^{\circ}.985$, and thus Venus gains daily on the earth by an amount equal to $1^{\circ}.612 - 0^{\circ}.985 = 0^{\circ}.627$. Let S V E be the line of the first

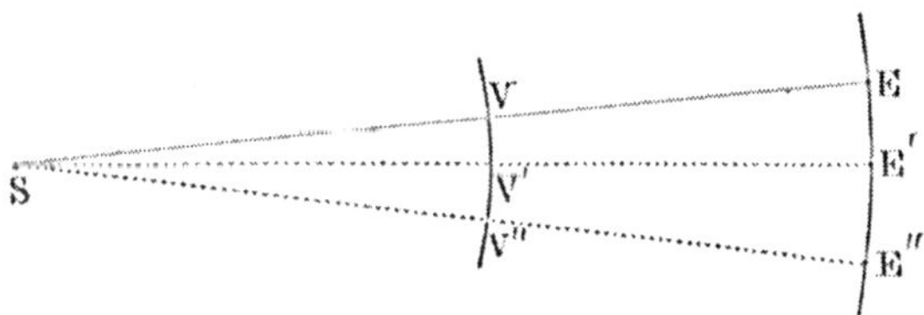

conjunction. At the end of thirteen revolutions of Venus she returns to the point V, while the earth is in the point E', requiring yet 0.852 days to reach E. Venus must, therefore, have passed the earth on some line as S V'' E'', such that Venus will have gained 0.852 days on the earth when she arrives at V. But the daily motion of Venus is $1^{\circ}.612$, and in the fraction of one day 0.852, she will move $1^{\circ}.612 \times 0.852 = 1^{\circ}.373$. Hence, the new line of conjunction S V″ E″ must fall behind the old line by this amount in each great cycle of thirteen revolutions of Venus and eight revolutions of the earth.

At the end of a great cycle, formed by dividing 360° by 1°.373, and multiplying the quotient by 8, the line of conjunction will return to its former position; and in case the orbits remain circular, all the perturbations of both the planets resulting from this cause, as affecting the orbital velocities and consequently the lengths of the major axes, will have been completely obliterated. The orbits are, however, not exact circles, neither are the elements invariable, and hence the restoration will not be perfect even at the end of this great cycle; but as the changes are all periodical, and as the lines of apsides revolve entirely around, periodicity again marks these minute perturbations, and at the end of a grand cycle, composed of many subordinate ones, these complexities and modifications will all be entirely swept away, and the system return to its primitive condition. The singular equation (as it is called) above described in the mean motions of Venus and the earth was first detected by the present Astronomer Royal. The period is about 240 years, and in the whole of this time the accumulated effect on the longitude of Venus cannot exceed 2″.95, while its effect on that of the earth only reaches 2″.06. This result of computation yet remains to be verified by actual observation.

For other particulars of the characteristics of this planet and of the elements of its orbit we refer the reader to the Appendix.

THE EARTH AND MOON AS PONDERABLE BODIES.—We have already determined the weight of the earth in terms of that of the sun, and we have seen that it would require 354,936 earths like ours to balance the ponderous orb which occupies the focal point of the solar system. It remains now to determine the absolute weight of the earth in pounds avoirdupois. We shall assume water as the standard, and admit that one cubic foot of water weighs 62.3211 lbs. From the known magnitude of the sphere of the earth, assuming, say, the *mean* diameter to be 7,941.12 miles, we can obtain the solid contents in cubic miles, amounting to no less than 259,373 millions. The number of cubic feet in a cubic mile is readily computed, being equal to $5{,}280 \times 5{,}280 \times 5{,}280$. Thus if we knew the weight of one cubic foot of the earth, in terms of the weight of one cubic foot of water, the total weight of the entire globe could be readily obtained in pounds.

This weighing of the earth, absolutely, is a problem of great difficulty, yet it has been executed, and the final results, though not precisely accurate, are, no doubt, close approximations. We can only give a general outline of the principle involved in the method employed. Suppose an inflexible rod with a small leaden ball at each extremity suspended in the middle by a delicate wire. When absolutely at rest the wire will hang vertically without twisting or torsion. Any force applied to either leaden ball to move it horizontally will tend to twist the suspending wire, and this torsion will resist the action of the force, and this resistance will finally be brought into equilibrium with the force, and will thus in some sense become its measure. Thus in case a delicate weight is attached to one of the leaden balls and suspended over a pulley, it will descend until the torsion of the wire shall be such as to exactly balance the small weight, and then the torsion and weight will stand in equilibrio, and the value of the weight (friction out of consideration) measures the force of resistance to torsion. Suppose a divided scale placed beneath the leaden ball, and a needle used as a pointer, then as the ball moves over this scale, a microscope properly adjusted may read the amount of motion with the greatest delicacy.

This machinery being arranged, suppose we bring a leaden ball one foot in diameter to within, say six inches of the small ball. Its power of attraction will move this ball over a space easily read off from the divided scale, and this will measure the attractive force of the large leaden ball.

Having thus learned the power exerted by a leaden ball one foot in diameter, on a material point located one foot from its center, it is easy, from the principles already laid down, to compute what would be the attractive power of a globe of lead as large as the earth; and in case this power of attraction thus computed should be precisely equal to that exerted by the earth, then the earth must weigh exactly as much as the leaden globe of equal size. This, however, is found from many experiments, tried with the most refined apparatus, not to be the case. The leaden globe is much heavier than the earthen one, and, indeed, we find that one cubic foot of earth of the mean density of the whole globe is as heavy as about five and a half cubic feet of water. Hence, every cubic foot of the earth weighs on the average $62.3211 \times 5.5 = 342.76$ lbs.; and as there are in the entire globe 259,373 millions of cubic feet, the whole globe must weigh $342.76 \times 259{,}373$ millions of pounds avoirdupois.

With this knowledge of the absolute weight of our

earth it is easy to obtain the weight of the sun and planets in pounds, were it necessary. Multiply the number of pounds in the weight of the earth by 354,936, and we obtain the actual weight of the sun.

THE FIGURE OF THE EARTH.—We have already seen that the earth is not a sphere, but a spheroid, protuberant at the equator and flattened at the poles. The exact methods of astronomy employed in the measurement of arcs of the earth's meridians, together with the vibrations of the pendulum in different latitudes, have fixed with great accuracy the relative values of the polar and equatorial diameters of the earth. The mean of a large number of measures results in giving the—

Equatorial diameter, ...	41,847,192 feet.
Polar diameter, ...	41,707,324 "
This gives a compression of	139,768 "

Some very remarkable results flow from this peculiar figure of the earth. Among these we shall consider first the *equilibrium of the ocean*. If it be true, as just asserted, that the equatorial diameter of our globe exceeds the polar diameter by 139,768 feet, then in case the earth were reduced to the figure of an exact sphere, by turning off the redundant matter, we should be compelled to turn down at the equator, to a depth of no less than 69,884 feet (one-half of the above quantity), and hence the equatorial region may be considered as a vast mountain range, belting the whole earth, and rising above the general level nearly seventy thousand feet. On the sides and over the summit of this mountain range the ocean sweeps its currents and its tides, and yet the most delicate and beautiful equilibrium is maintained.

This is due to the fact that the velocity of rotation of

the earth on its axis is absolutely uniform and invariable, and hence the *centrifugal force*, whose power precisely counterbalances the gravity on the mountain side on which the ocean rests, is ever the same. This great principle is beautifully exemplified by taking a glass vase, filling it with a colored liquid, and suspending it by a cord. So long as the vase is at rest the fluid on its upper surface is precisely level and plain. Now, give to the vase a motion of rotation about a vertical axis (as by the untwisting of the suspending cord), and at once the fluid commences to rise upon the sides of the vase, and a disk-shaped cavity is formed. This rising continues so long as the velocity of rotation increases. Should the velocity become uniform, then the figure of the fluid in the vase assumes a form of exact equilibrium, and the delicate circle that marks the height to which the fluid rises in the vase remains constant, and will so continue, so long as the velocity of rotation is unchanged. The stability of the figure of the ocean depends on the same principle, and were it possible to arrest the rotation of the earth, instantly the equatorial ocean would rush towards the poles and would there rise until the general level should become such as is due to a spherical figure, which the ocean would assume.

Nutation and precession.—We have in these remarkable phenomena another effect of the figure of the earth. We have already mentioned the fact that the *vernal equinox* (the point in which the sun's center crosses the equinoctial in the spring season), does not remain fixed in the heavens. This discovery was made by the early astronomers, the fact noted, and an approximate period of revolution, amounting to some twenty-five or twenty-six thousand years. Modern science has not only determined the exact period to be 25,868 years, but has traced the phenomenon to its origin, and has revealed the cause to lie in the fact that the protuberant mass of matter surrounding the equator of the earth is a sufficient *purchase* to enable the sun and moon to *tilt* the entire earth, and consequently the plane of the earth's equator. Suppose the earth revolved on an axis perpendicular of the ecliptic, or that the equator of the earth and the ecliptic coincided. Then, so far as the sun is concerned, there could arise no power to effect a change in the plane of the equator; but as the moon revolves in an orbit inclined to the plane of the ecliptic, the moon will be sometimes above and sometimes below the plane of the earth's equator, now supposed to be coincident with the ecliptic. Whenever the moon is above the equatorial ring of the earth, she will tend to lift the nearest portion of that ring above the ecliptic, and to sink the opposite part below the same plane. and as the moon revolves around the earth, she will cause the equatorial ring to tilt *towards* her position, and thus the *line of nodes* of the ring will revolve as do the lines of nodes of the planetary orbits; and this is precisely what we find to be true of the line of equinoxes, or the line cut by the plane of this equatorial ring from the plane of the ecliptic, producing, as we have seen, a *retrocession* of the equinoctial point, and a precession of the time of the equinox.

THE NUTATION OF THE EARTH'S AXIS is a phenomenon springing out of the same causes producing precession. If we consider the axis of the earth as an inflexible bar, passing through the earth's center and perpendicular to the equator, extending indefinitely in opposite directions, to the celestial sphere, it is clear that any tilting of the earth's equatorial ring will equally tilt the axis of the earth. This is actually seen in the slow revolution of the pole of the earth's equator around the pole of the ecliptic in a period precisely equal to that employed in the revolution of the equinoxes. Nutation is but a subordinate fluctuation whereby the pole of the equator, instead of describing an exact circle around the pole of the ecliptic, makes certain short excursions a little on the inside and on the outside of this circle, in a period which agrees exactly with that occupied by the revolution of the nodes of the moon's orbit. This at once suggests the moon to be the principal cause of this *nodding* of the earth's axis, and, indeed, modern analysis has pointed out the origin of the movement, and has accurately computed its value.

In all we have said we have supposed the equator and ecliptic to coincide. This, however, is not the case of nature. These planes are inclined to each other, and hence we find the sun producing results (analogous to those already traced to the moon) on the mass of protuberant matter surrounding the earth's equator. The exact values of these *constants* of *precession* and *nutation* will be found in the Appendix. The greatest pains have been bestowed on their determination, as they are of the first importance in fixing the absolute places of all the heavenly bodies.

Figure of the earth's orbit.—The ellipticity of the earth's orbit is slowly wearing away, under the combined influence of all the planets. The eccentricity at the commencement of the present century amounted to 0.016783568, the semi-major axis being considered as unity. The amount by which this quantity is decreased in a hundred years is 0.00004163. Let us reduce these figures to intelligible quantities. The eccentricity is the distance from the center of the ellipse to the focus, and in miles is equal to $0.016783568 \times 95,000,000 = 1,594,100$. This quantity decreases in one hundred years by $0.00004163 \times 95,000,000 = 3,954.85$ miles. If, now, we divide 1,594,100 by 3,954.85, the quotient 405 + will be the number of centuries which must elapse before the earth's orbit will become an exact circle at the present rate of change. It is ascertained by a rigorous analytical investigation of this great problem that so soon as the circular figure is reached by the earth's orbit the same causes reverse their effects, and the circular figure is lost, and the eccentricity of the elliptic figure slowly increases until finally, at the end of a vast period, the original form of the orbit is regained, to be again lost, and thus an expansion and contraction marks the history of the earth's orbit, vibrating through periods of time swelling into millions of years.

Acceleration of the moon's mean motion.—This change in the figure of the earth's orbit produces a minute change in the mean motion of the moon, which was, after long years of the most laborious research, finally traced to its true origin by La Place. The fact that the moon was moving faster in modern than in ancient times became evident from a comparison of the modern and ancient eclipses. These eclipses can only occur when the sun, earth and moon occupy the same right line nearly, and hence their record gives a very precise knowledge of the relative position of these three bodies. It thus became manifest that the average speed with which the moon was moving in her orbit was slowly increasing from century to century. This follows necessarily from the fact that the loss of eccentricity by the orbit removes the earth by a small amount (on the average) further from the sun. This carries both the earth and her satellite by so much away from the disturbing influence of the sun, leaving to the earth a more exclusive control of the moon. As the sun is outside the moon's orbit with reference to the earth, his attraction will increase the magnitude of the moon's orbit, and, of course, her periodic time. Any diminution of the sun's disturbing power will therefore by so much permit the moon to approach the earth, and to increase her velocity of revolution; and this is precisely what observation has revealed with reference to our satellite during the entire period that history has recorded the progress of astronomy.

This gradual acceleration must continue up to the time when the earth's orbit shall become exactly circular in form. This limit once attained, as this orbit slowly resumes its elliptic form, the acceleration of the moon's mean motion is converted into retardation, and thus at the end of a mighty period this change will be entirely destroyed, and the moon and earth return to their primitive condition. This acceleration of the mean motion of the moon is so slow that from the earliest record of eclipses by the Babylonians down to the present time, some 2,500 years, the moon has got in advance of her mean place by about three times her own diameter.

The facts above related indicate with how much diligence the moon's motions have been studied. Though she is our nearest neighbor, and consequently more directly under the eye of the astronomer than any other heavenly body, her motions have been more complex and difficult of perfect exposition than any object in the heavens. The recent investigations of the European and

American astronomers and mathematicians seem to have finally conquered this refractory satellite, so that now it becomes possible to unravel her involved and intricate march among the stars with such precision that we can fix her place with certainty for even thousands of years.

MARS AS A PONDERABLE BODY.—This planet revolves in an orbit of such eccentricity as to present very marked differences in the power of attraction of the sun on the planet when at its greatest and least distances. Its mean distance is 142 millions of miles, giving its semi-major axis a length equal to 71 millions of miles. The eccentricity of the orbit (the distance between the center and focus of the ellipse) amounts to nearly one-tenth of this quantity, or to about 6.4 millions of miles. Hence, the perihelion distance is 64.6 millions of miles, while the aphelion distance is 77.4 millions of miles. The attractive power exerted by the sun in perihelion will be greater than that exerted in aphelion in the ratio of $(77.4)^2$ to $(64.6)^2$, or as 5,991 to 4,172, or nearly as 3 to 2. To resist this increased power of attraction in perihelio the planet must there move with a far higher velocity than when in its aphelion. All these deductions from theory are verified by observation.

It was from an examination of the movements of Mars that Kepler deduced his celebrated laws. These laws we have had occasion to use constantly in our computations, but in consequence of the mutual actions of the planets, not one of these laws is rigorously true. The orbits of the planets are not exact ellipses, nor do they so revolve that the lines joining them with the sun sweep over precise equal spaces in equal times, nor are the squares of the periods of revolution precisely proportional to the cubes of the mean distances; but the failure in these laws is due entirely to mere *perturbation*, and in case a single planet existed revolving around the sun, they would all be scrupulously fulfilled.

The planet Mars was, however, well situated for the examination conducted by Kepler. This becomes manifest if we call to mind the great distance separating Mars and Jupiter, and the comparatively small disturbance which the earth can produce. To present this problem still clearer let us suppose the earth, Mars and Jupiter to be in conjunction, and situated as in the figure. Then

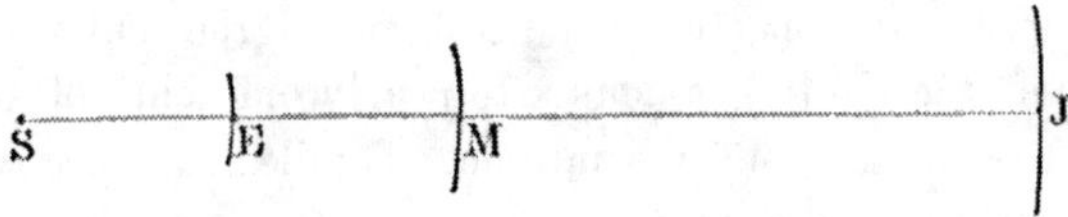

the distance from S to E is 95 millions of miles, from S to M 142 millions, from S to J 890 millions of miles. Hence, the distance E M is 142−95=47 millions of miles, while the distance M J is 890−142=648 millions of miles. We will first compute the power of attraction of Jupiter on Mars, as compared with the power of the sun. If the masses were equal the energy of Jupiter would, on account of the greater distance, be reduced below that of the sun in the ratio of $(142)^2$ to $(648)^2$, or nearly as 1 to 21. But the masses are not equal, for the sun weighs as much as 3,502 such globes as Jupiter, and hence, by combining these causes of reduction, we find the force exerted by Jupiter to be less than that exerted by the sun in the ratio of 1 to 3,502×21, or as 1 to 115,542.

Let us now see what force the earth exerts on Mars, when compared with the sun's force. As the earth is nearer to Mars than the sun, in case the sun and earth were of equal weights their energy at Mars would be in the ratio of $(142)^2$ to $(47)^2$, or as 20,164 to 2,209, or nearly in the ratio of 9 to 1. But the sun weighs as much as 354,936 earths, and if we divide 354,936 by 9, we obtain 39,437 as a quotient, and hence the power of the sun on Mars is to the power of the earth as 39,437 to 1.

It is thus seen that the earth is more powerful than Jupiter to disturb Mars in the ratio of 115,542 to 39,437, or in the ratio of about 3 to 1. To exhibit more clearly the minute character of the effects of the earth and of Jupiter on this planet, let us compute the space through which a body would fall in one second, if as far removed from the sun as Mars. We have already seen that gravity at the solar surface is 28.7 greater than at the surface of the earth. At the earth gravity impresses such a velocity on a falling body that it passes over a space of 16.1 feet in the first second of time; therefore, a body at the sun's surface would fall through a space represented by $28.7 \times 16.1 = 461$ feet in the first second of its fall. If we remove the falling body to double the distance from the sun's center, the force of the sun's gravity is reduced to one quarter, and hence the space passed over by the falling body at two units from the center of the sun will be $\frac{461}{4} = 115.25$ feet. But Mars' distance from the sun is 142 millions of miles, while the solar radius is 441,500 miles; in other words, a falling body, removed to the distance of Mars from the sun, is about 32 times more remote from the sun's center than when on the sun's surface, and the energy of the sun's gravity would be reduced at this distance in the ratio of $(1)^2$ to $(32)^2$, or as 1 to 1,024; so that a body would fall in one second, if as far removed from the center of the sun as is the planet Mars, through a space represented by $\frac{461}{1024}=0.450194$ feet. To what extent will this quantity be affected by the attraction of the earth? The answer is given in the result already reached that the power of the earth is only the thirty-nine thousand four hundred and thirty-seventh part of that of the sun, and hence the falling body will only pass over the additional space represented by the minute fraction $\frac{0.450194}{39437}=$ 0.000011, or about the one hundred thousandth part of one foot. These quantities look to be minute and quite unworthy of notice, and yet from these small disturbing effects, accumulating through ages, arise all the amazing changes which are progressing among the elements of the planetary orbits.

We will not extend these details, but refer for further particulars to the Appendix.

THE ASTEROIDS AS PONDERABLE BODIES.—As yet we have no certain knowledge of the magnitude or masses of these minute worlds. We are assured that they are subjected to the laws of motion and gravitation, and that the elements of their orbits are undergoing the same modifications to which the elements of the orbits of all the planets are subjected. These planets are disturbed principally by the action of Jupiter, as we may readily determine by an examination of the masses and distances of the two nearest planets, inside and outside the orbits of the asteroids.

The mean distance of the group from the sun is about 2.5 times the earth's distance. The distance of Mars, in terms of the same unit, is 1.5, and the distance of Jupiter from the sun is 5.2. Hence, from the mean distance of the asteroids to Jupiter is $5.2-2.5=2.7$, and from the same to Mars is 2.5—1.5=1.0. Hence, if Mars and Jupiter were equal in weight, the power of Mars over the central asteroid would exceed the power of Jupiter in the ratio of $(2.7)^2$ to $(1.0)^2$, or in the ratio of 7.3 to 1. But Jupiter is 256 times heavier than Mars, and hence his power will be increased in like proportion, and the attraction of Mars will be to that of Jupiter as 7.3 to 256, or as 1 to 35 nearly. Hence we perceive that Jupiter is the principal disturber in the movements of the asteroids. For further particulars the reader will consult the Appendix.

JUPITER AND HIS SATELLITES AS HEAVY BODIES.—This planet is not only heavier, but its volume is much greater than that of any one of the planets. Being 5.2 further from the sun than the earth, it will be attracted by a power diminished in the ratio of the square of 5.2 to 1, or as 27 to 1 nearly.

The weight of Jupiter is to that of the earth as 338 to 1, and in case his diameter were exactly equal to that of the earth, a body weighing one pound at the terrestrial equator would weigh at the equator of this planet 338 pounds. But the diameter of Jupiter is 90,734 miles, and its radius 45,377 miles, or more than ten times the radius of the earth. His attraction on a body upon the surface will therefore be reduced on account of this tenfold distance to the one hundredth of 338 pounds, or to 3.38 pounds, or, if the computation be made precisely, the result gives us 2.81 as the weight of one terrestrial pound at the equator of this planet. The student can compute the reduction in the gravity of the planet at the equator arising from the action of the centrifugal force, the planet revolving on its axis in 9h. 55m. 27s.

The principal disturber of Jupiter is the planet Saturn.

From the sun to Jupiter is 5.2, the earth's distance being 1. From Jupiter to Saturn the distance is 4.3 in the same terms. Hence, if Saturn weighed as much as the sun, his power over Jupiter would be greater in the ratio of $(5.2)^2$ to $(4.3)^2$, or as 27.04 to 18.44, or as 1.47 to 1. But the sun weighs as much as 3,502 Saturns, and hence his power over Jupiter will exceed that of Saturn in the ratio of $\frac{3502}{1.47}$ to 1, or as 2,380 to 1. This, of course, is the ratio of the forces when the planets are in conjunction. When in opposition the interval between them is increased by the diameter of the orbit of Jupiter, or by 10.4, and thus it becomes 14.7, instead of 4.3, and in this position the disturbing power of Saturn is reduced in the ratio of $(14.7)^2$ to $(4.3)^2$, or as 216.09 to 18.44, or as 12 to 1.

We have already seen how we can ascertain the weight of this planet by observing the period of revolution of his satellites and by measuring their distances. By taking these quantities from the table in the Appendix the student may compute readily the mass of Jupiter as compared with that of the earth.

The eccentricity of the orbit of this planet amounts to 0.0481, the semi-major axis being unity, or the distance from the center of the ellipse to the focus is equal to $(242,500,000) \times 0.0481 = 11,664,250$ miles. This quantity is now slowly increasing, and gains every year in length 388 miles. This is due to the disturbing influence of the surrounding planets, and after an immense period will reach a limit beyond which it cannot pass. The increment will then be converted into decrement, and a limit being again reached, the orbit in its figure thus oscillates between these limits in calculable, but (so far as I know) in periods not yet calculated.

The same fact is true of the inclination of the plane of Jupiter's orbit to that of the ecliptic. On the 1st January, 1840, this inclination amounted to 1° 18′ 42″.4. Its present annual decrease is 0″.23, and should this continue, at the end of about 200,000 years these planes would coincide. This, however, can never take place. The decrease finally comes to be converted into increase, and thus the plane of the orbit of Jupiter may be said to rock to and fro on the plane of the ecliptic in periods reaching to even millions of years.

We have already noticed a source of perturbation in the case of Venus and the earth, arising from the approximate commensurability of the periods of revolution of these planets. A like *equation*, as it is called, exists in the case of Jupiter and Saturn. Five periods of Jupiter are 21,663, and two of Saturn's periods are 21,519 days; so that in case the planets start at any given time from a conjunction, at the end of five revolutions of Jupiter and two of Saturn, the planets will return to nearly the same points of their orbits and to the same relative positions. But the synodical period, or the time from conjunction to conjunction of these planets is 7,253.4 days, and three times this quantity amounts to 21,760.2. Hence we perceive Jupiter in this period will have performed five revolutions and 21,760 – 21,663 = 97 days over, while Saturn will describe two revolutions and 240 days over, and during these excesses the planets advance in their respective orbits 8° 6′. Thus every third conjunction will fall 8° 6′ in advance of the former one, and the conjunction line will be thus carried round the entire orbit in about 44 times × 21,760 days, or in 2,648 years, at the end of which cycle the same exact condition will be restored, and all the perturbations in the same time completely obliterated, provided the figures of the orbits remain unchanged. Indeed, a restoration is effected partially and almost completely in consequence of the triple conjunction which takes place in the period of 21,760 days. These conjunctions fall at points on the orbits 120° apart, and thus tend to effect a restoration, which is only fully perfected, however, at the end of the great cycle of 2,648 years.

Here we find again the cause which prevents the laws of Kepler from being rigorously applicable to the planetary movements. In case Jupiter existed alone, then the line drawn from the planet to the sun would sweep over equal areas in equal times, as it is carried by the planet around the sun. But the association of the two planets renders the application of this law no longer possible. Jupiter is dragged back by Saturn, and Saturn is dragged forward by Jupiter when they start off from their line of conjunction; but here comes in a most wonderful compensation in the fact that whatever Jupiter's motion *loses* by the disturbing influence of Saturn, Saturn's motion gains by the disturbing influence of Jupiter. So that the *sum* of the areas swept over by the lines joining the two planets with the sun *will always be equal in equal times.*

We shall not extend further our notices of the results arising from the action of gravitation on the planets and their satellites—having discussed to some extent the mutual perturbations of Uranus and Neptune in a former chapter.

We will close by an extension of the principle laid down in the case of Jupiter and Saturn to the entire planetary system. If at any moment lines were drawn from the center of the sun to each of the planets in the entire system, and from the center of each of the planets to their respective satellites, the areas swept over by all these lines thus drawn will always be equal in equal times. Thus, while not a solitary planet or satellite can follow this law of equal areas, the combined scheme is bound by it in the most rigorous manner; and if the amount of area described by the entire system in one hour were determined to-day, and be sent down to posterity, at the end of ten thousand years, a like computation being made, the same identical result will be reached, provided the system remain free from any disturbing influence exterior to itself.

In case, therefore, the sun with all his planets and comets is, indeed, drifting through space into other stellar regions, the time may come when the fixed stars may so disturb the *sum of the areas* as to point out clearly the fact that our system has positively changed its location in space.

We will close our discussion of the sun and his satellites by the examination of an hypothesis which has been propounded to account for the peculiar organization of this vast scheme of revolving worlds.

CHAPTER XVI.

THE NEBULAR HYPOTHESIS.

THE ARRANGEMENT OF THE SOLAR SYSTEM.—THE PHENOMENA FOR WHICH GRAVITATION IS RESPONSIBLE.—THE PHENOMENA REMAINING TO BE ACCOUNTED FOR.—NEBULOUS MATTER AS FOUND IN COMETS.—NEBULOUS MATTER POSSIBLY IN THE HEAVENS.—THE ENTIRE SOLAR SYSTEM ONCE A GLOBE OF NEBULOUS MATTER.—MOTION OF ROTATION.—RADIATION OF HEAT.—CONDENSATION AND ITS EFFECTS.—RINGS DISENGAGED FROM THE EQUATOR OF THE REVOLVING MASS.—FORMATION OF PLANETS AND OF SATELLITES.

IN our examination of the scheme of worlds which revolve around the sun we have found that the orbits of the planets are all nearly circular, that their planes are all nearly coincident with the plane of the ecliptic, and that this plane is nearly coincident with the plane of the sun's equator; that the planets all revolve in the same direction around the sun, and that the sun and planets and satellites all rotate on their axes in the same direction; that the periods of revolution grow shorter in the planets and satellites as their distances from their primary grows less; that the sun rotates on his axis in a shorter period than that employed in the revolution of any planet; that every planet accompanied by satellites rotates on its axis in a less time than the period of revolution of any satellite. The law of gravitation is not responsible for any of these facts, and in case we compute the chances of such an organization coming into being by accident, we shall find but one chance in so many millions that we are compelled to look to some higher cause than mere accident to account for so great a multitude of combined phenomena.

We have said that gravitation is not responsible for the facts above stated. In case a solitary planet be projected with a given force, and in a given direction about the sun, and at a given distance, it will revolve, as we have seen, in one of four curves, and in any one of these curves it will be held equally by the law of gravitation. The plane in which it revolves may assume any angle with a fixed plane, the direction of the revolution may be the same or contrary to that in which the sun rotates, the orbit may be a circle, an ellipse, a parabola, or an hyperbola, and yet the planet shall revolve, subject to the law of gravitation. It may rotate on its own axis either with or against its revolution in its orbit, and in case we give to this planet a satellite, the same statements are true with reference to this attendant. So that, so far as the law of gravitation is concerned, there might have been among the planets all the diversity in the form of their orbits in the angles of their inclination to a fixed plane, and in the direction of their motions, as are found among the comets, and yet each object would have been subject to the great law of universal gravitation.

We cannot, therefore, affirm that the peculiar structure of the solar system results from the laws of motion and gravitation, without pre-supposing a condition of matter entirely different from that now recognized as existing in the planets and their satellites. We have already noticed the wonderful constitution of the *comets*. In these bodies is found a kind of matter which has been termed *nebulous*, in which the minute particles are separated by some repulsive force, and the entire mass is but a vapor of the most refined tenuity.

Among the stellar regions the telescope has revealed objects whose light is so faint and whose forms are so ill defined that they have been regarded by many astronomers of high reputation to be analogous to the comets in their material, exhibiting the primitive or primordial condition of the matter composing the physical universe. This conjecture (for it is nothing more) may be true or false, but its truth or falsehood cannot in any way affect the credibility of the theory or hypothesis we are about to present. The present condition of matter cannot in any way be assumed to be the only condition in which it ever existed, since we now know it to be subject to extraordinary changes and most wonderful modifications.

Let us, then, suppose that a time once was when the sun and all its planets and their satellites existed as one mighty globe of nebulous matter, whose diameter far exceeded the present diameter of the orbit of Neptune, that to this stupendous globe a motion of rotation was given, and that its heat is slowly lost by radiation, and let us endeavor to follow the changes which must flow from the loss of heat and the operation of the laws of motion and gravitation, and learn whether from this parent mass a scheme of planets and satellites such as now exist can be generated. We prefer to present the reasoning in the language of M. Pontecoulent, one of the most eminent of the illustrious disciples of Newton—merely premising that in case the central rotating mass contracts by loss of heat that a time must come when, in consequence of the increased velocity of rotation, the force of gravity of a particle at the equator will be overcome by the centrifugal force generated by the velocity of rotation, and hence flat zones or rings of vapor or nebulous matter must eventually be formed in the plane of the equator of the revolving globe :—

"These zones must have begun by circulating round the sun in the form of concentric rings, the most volatile molecules of which have formed the superior part, and the most condensed the inferior part. If all the nebulous molecules of which these rings are composed had continued to cool without disuniting, they would have ended by forming a liquid or solid ring. But the regular constitution which all parts of the ring would require for that, and which they would have needed to preserve whilst cooling, would make this phenomenon extremely rare. Accordingly the solar system presents only one instance of this, that of the rings of Saturn. Generally the ring must have broken into several parts, which have continued to circulate round the sun, and with almost equal velocity, while at the same time, in consequence of their separation, they would acquire a rotatory motion round their respective centers of gravity; and as the molecules of the superior part of the ring, that is to say, those furthest from the center of the sun, had necessarily an absolute velocity greater than the molecules of the inferior part which is nearest it, the rotatory motion, common to all the fragments, must always have been in the same direction as the orbitual motion.

"However, if after their division one of these fragments has been sufficiently superior to the others to unite them to it by its attraction, they will have formed only a mass of vapor, which, by the continual friction of all its parts, must have assumed the form of a spheroid flattened at the poles and elongated in the direction of its equator. Here, then, are rings of vapor left by the successive retreats of the atmosphere of the sun, changed into so many planets in the condition of vapor circulating round the sun, and possessing a rotatory motion in the direction of their revolution. This must have been the most common case; but that in which the fragments of some ring would form several distinct planets possessing degrees of velocity must also have taken place, and the telescopic planets discovered during the present century seem to present an instance of this; at least if it is not admitted with Olbers, that they are the fragments of a single planet, broken by a strong interior commotion. It is easy to imagine the successive changes produced by cooling on the planets whose formation has been just pointed out. Indeed, each of these planets, in the condition of vapor, is, in every respect, like one of the nebula in the first stage; they must, therefore, before arriving at a state of solidity, pass through all the stages of change we have just traced in the sun. At first the condensation of their atmosphere will form round the center of the planet a body composed of layers of unequal density, the densest matter having, by its weight, approached the center, and the most volatile reached the surface, as we see in a vessel different liquids ranged one above another, according to their specific gravity to arrive at a state of equilibrium. The atmosphere of each planet will, like that of the sun, leave behind it zones of vapor, which will form one or several secondary planets, circulating round the principal planet as the moon does round the earth, and the satellites round Jupiter, Saturn, and Uranus, or else they will form, by cooling without dividing, a solid and continuous circle, of which we have an instance in the ring of Saturn. In every case the direction of the rotatory and orbital motion of the satellites or the ring will be the same as that of the rotatory motion of the planet; and this is completely confirmed by observation.

"The wonderful coincidence of all the planetary motions, (a phenomenon which we cannot, without infringing the laws of probability, regard as merely the effect of chance,) must then be the result even of the formation of the solar system on this ingenious hypothesis; we see also why the orbits of the planets and satellites are so little eccentric, and deviate so little from the plane of the solar equator. A perfect harmony between the density and temperature of their molecules in a state of vapor would have rendered the orbits rigorously circular and made to coincide with the plane of this equator; but this regularity could not exist in all parts of such large masses; there has resulted the slight eccentricities of the orbits of the planets and satellites, and their deviation from the plane of the solar equator.

"When in the zones abandoned by the solar atmosphere there are found molecules too volatile either to unite with each other or with the planets, they must continue to revolve round the sun, without offering any sensible resistance to the motions of the planetary bodies, either on account of their extreme rarity, or because their motion is effected in the same way as that of the bodies they encounter. These wandering molecules must thus present all the appearances of the zodiacal light.

"We have seen that the figure of the heavenly bodies was the necessary result of their fluidity at the beginning of time. The singular phenomenon presented by the rigorous equality indicated by observation among the lesser motions of rotation and revolution of each satellite, an equality rendering the opposed hemisphere of the moon forever invisible to us, is another obvious consequence of this hypothesis. Indeed, supposing that the slightest difference had existed between the mean motion of rotation and revolution of our satellite while it was in the state of vapor or of fluidity, the attraction of the earth would have elongated the lunar spheroid in the direction of its axis towards the earth. The same attractions would have tended to diminish insensibly the difference between the rotatory and orbitual motions of the moon, so as to confine to narrow limits a condition sufficient to cause the axis of its equator, directed towards the earth, to be subject only to a species of periodical balancing constituting the phenomenon of libration. If these oscillations are not now observed, it is because they have ceased to exist in consequence of the resistance they have encountered in the course of time, even as the oscillations of the terrestrial axis in the interior of the earth, arising from the initial state of motion, have been destroyed, and as indeed all the motions of the heavenly bodies have disappeared which have not had a permanent cause.

"The principal phenomena of the planetary system are therefore explained with great facility by the hypothesis we are examining; and as these successive changes of a nebulous mass and the leaving of a part of its substance by cooling, agree with all the leading phenomena, it must be allowed a high degree of probability. In this hypothesis the formation of the planets would not have been simultaneous; they have been created successively at intervals of ages; the oldest are those which are furthest from the sun, and the satellites are of a more recent date than their respective planets. It may be, if we are ever permitted to reach so high, that by an examination of the constitution of each planet we may go back to the epoch of its formation, and assign to each its place in the chronology of the universe. It is likewise seen that the velocity of the orbitual motion of each planet, as it is now, must differ little from that of the rotatory motion of the sun at the period when the planet was detached from its atmosphere. And as the rotatory motion is accelerated in proportion as the solar molecules are confined by cooling, so that the sum of the areas which they describe round the center of gravity would remain always the same, it follows revolutionary motion must be so much more rapid as the planet is nearer the sun, as is seen by observation. It likewise results that the duration of the rotation either of the sun or of a planet must be shorter than the duration of the nearest body which circulates round them; this observation is completely confirmed even in those cases where the difference between the duration of the two motions must be very slight. Thus the interior ring of Saturn being very close to the planet, the duration of its rotation must be almost equal, but a little longer than that of the planet.

"The observations of Herschel give, indeed, 0.432 as the duration of the rotation of the ring, and 0.427 as that of the planet; why, then, should we not admit that this ring has been formed by the condensation of the atmosphere of Saturn, which formerly extended to it? We may perhaps deduce from the laws of mechanics and the actual dimensions of the sun, and the known duration of its rotation, the relation existing between the radius vector of its surface and the time of its rotation in the different stages of concentration through which it has passed. The third law of Kepler would be no longer the mere result of observation; it would be directly deduced from the primordial laws of the heavenly bodies.

"In this system the particular form of the planets, the flattening at the poles, and bulging out at the equator, is only the necessary consequence of the laws of the equilibrium of fluids, and easily explains the greater part of the phenomena observed by geologists in the constitution of the terrestrial globe, which appear inexplicable, if it is not admitted that the earth and planets have been originally fluid.

"Let us now see what is the origin and part assigned to comets by this hypothesis. La Place supposes that they do not belong to the planetary system, and he regards them as masses of vapor formed by the agglomeration of the luminous matter diffused in all parts of the universe, and wandering by chance in the various solar systems. Comets would thus be, in relation to the planetary system, what the aerolites are in relation to the earth, with which they seem to have no original connection. When a comet approaches sufficiently near the regions of space occupied by our system to enter into the sphere of the sun's influence, the attraction of that luminary, combined with the velocity acquired by the comet, causes it to describe an elliptic or hyperbolic orbit. But as the direction of this velocity is quite arbitrary, comets must move in every direction and in every part of the sky.

"The cometary orbits will, then, have every inclination to the ecliptic; and this hypothesis explains equally well the great eccentricity by which they are usually effected. Indeed, if the curves described by comets are ellipses, they must be greatly elongated, since their major axes are at least equal to the radius of the sphere of the sun's attraction; and we must consequently be able to see only those whose eccentricity is very great, and perihelion distance inconsiderable; all others, on account of their minuteness and distance, must always be invisible, unless at least the resistance of the ether, the attraction of the planets, or other unknown causes diminish their perihelion distance, and bring them nearer the terrestrial orbit. The same circumstances may change the primitive orbits of some comets into ellipses, whose major axes are comparatively small; and this has probably happened to the periodical comets of 1759, 1819, and 1832. The laws of the curvilinear motion likewise show that the eccentricity of the orbit chiefly depends on the direction of the comet's motion on its entering the sphere of the sun's attraction; and as this motion is possible in every direction, there are no limits to the eccentricities of the orbits of comets.

"If, at the formation of the planets, some comets penetrated the atmospheres of the sun and planets, the resistance they met would gradually destroy their velocity; they would then fall on those bodies describing spirals, and their fall would have the effect of causing the planes of the orbits and equators of the planets to remove from the plane of the solar equator. It is, therefore, partly to this cause, and partly to those we have developed above, that the slight deviations we now perceive must be attributed.

"Such is a summary of the hypothesis of La Place on the origin of the solar system. This hypothesis explains, in the most satisfactory manner, the three most remarkable phenomena presented by the planetary motions.

"1st The motion of the planets in the same direction, and nearly in the same plane.

"2d. The motion of the satellites in the same direction as their planets.

"3d. The singular coincidence in direction of the rotatory and orbitual motions of the planets and the sun, which in other systems would present inexplicable difficulties.

"The no less remarkable phenomena of the smallness of the eccentricities and inclinations of the planetary orbits are also a necessary consequence of it, while we see at the same time why the orbits of the comets depart from this general law, and may be very eccentric, and have any inclination whatever to the ecliptic. The flattening of the form of the planets, shown on the earth by the enlargement of degrees of the meridian, and by the regular increase of weight in going from the equator to the poles, is only the result of the attraction of their molecules while they were yet in a state of vapor, combined with the centrifugal force produced by the rotatory motion impressed on the fluid mass. In short, among the phenomena presented by the motions and the form of the heavenly bodies, there are none which cannot be explained with extreme facility by the successive condensation of the solar system; and the more this system is examined the more we are led to acknowledge its probability.

"Undoubtedly, if, as La Place has himself said, a hypothesis not founded on observation or calculation must always be presented with extreme diffidence, this, it will be granted, acquires, at least by the union and agreement of so many different facts, all the marks of probability. But what, in my opinion, principally distinguishes it from the ordinary theories concerning the formation of systems, is the identity which it establishes between the solar system and the stars spread so profusely through the sky.

"All the phenomena of nature are connected, all flow from a few simple and general laws, and the task of the man of genius consists in discovering those secret connections, those unknown relations which connect the phenomena which appear to the vulgar to have no analogy. In going from a phenomenon of which the primitive law is easily perceived, to another in which particular circumstances complicate it so as to conceal it from us, he sees them all flowing from the same source, and the secret of nature becomes his possession. Thus the laws of the elliptic motion of the planets led Newton to the great principle of universal gravitation, which he would have sought for in vain in the less simple phenomena of the rotatory motion of the earth, or the flux and reflux of the sea. But this great principle being once discovered, all the circumstances of the planetary motions were explained, even in their minutest details, and the stability of the solar system was itself only the necessary consequence of its conformation, without which, as Newton thought, God would be constantly obliged to retouch his work, in order to render it secure. La Place, extending to all the stars, and consequently to the sun, the mode of condensation by which the nebula are changed into stars, has connected the origin of the planetary system with the primordial laws of motion, without recurring to any hypothesis but that of attraction. He has, therefore, extended to the fixed stars the great law of universal gravitation, which is probably the only efficient principle of the creation of the physical world, as it is of its preservation."

Such is a brief outline of one of the most sublime speculations that has ever resulted from the efforts of human thought. It carries us back to that grand epoch when "in the beginning God created the heavens and the earth," when matter was first called into being in its unformed nebulous condition, and "the earth was without form and void," and darkness covered the mighty deep of unfathomable space. But the Spirit of God moved on the boundless flood of vaporous matter scattered through the dark profound, and gave to each particle its now eternal function, impressed the laws of gravitation and motion, selected the grand centers about which the germs of suns and systems should form, and in infinite wisdom drew the plan of that one scheme which we have attempted to examine, among the millions that shine in splendor throughout the boundless empire of space.

APPENDIX.

TABLES OF ELEMENTS.

SOLAR ELEMENTS, EPOCH 1st JAN., 1801.

Mean longitude	280° 39′ 10″.2
Longitude of the perigee	270° 30′ 05″.0
Greatest equation of center	1° 55′ 27″.3
Decrease of same in one year	0″.173
Inclination of axis to the ecliptic	82° 40′ 00″.0
Motion in a mean solar day	7° 30′ 00″.0
Motion of perigee in 365 days	1′ 01″.9
Apparent diameter	32′ 12″.6
Mean horizontal parallax	8″.6
Rotation in mean solar hours	607h 48m 0s
Time of passing over one degree of mean longitude	24h 20m 58.1s
Eccentricity of orbit (semi-axis major 1)	.01685318
Volume (earth as 1)	1415225
Mass (earth as 1)	354936
Mean distance in miles	95,000,000
Same (earth's radius 1)	23984
Density (earth as 1)	0.250
Diameter in miles	888,646
Gravity at equator	28.7
In one second of time bodies fall in feet	462.07

ELEMENTS OF THE ORBIT OF MERCURY, EPOCH 1st JAN., 1801.

Mean distance from the sun in miles	36,725,000
Same (earth's distance as 1)	.3870984
Greatest distance, same unit	.4666927
Least distance, same unit	.3075041
Eccentricity (semi-axis major as 1)	.2056178

Annual variation of same (decrease)	0.000,000,03866
Sidereal revolution in days	87.9692824
Synodical revolution in days	115.877
Longitude of perihelion at epoch	74° 57′ 27″.00
Annual variation of same (increase)	5.81
Longitude of ascending node	46° 23′ 55″.00
Annual variation of same	10″.07
Inclination of orbit to ecliptic	7° 00′ 13″.30
Annual variation of same	00″.18
Mean daily motion in orbit	245′ 32″.6
Time of rotation on axis	24h 05m 28s
Inclination of axis to ecliptic	Uncertain.
Apparent diameter	6″.69
Diameter in miles	3089
" (earth's being 1)	0.398
Volume (earth's being 1)	0.0595
Density (earth's being 1)	1.225
Light received at perihelion (earth's being 1)	10.58
Same at aphelion (earth's being 1)	4.59
Weight of a terrestrial pound	0.48
Space fallen through in one second of time, in feet	7.70
Mass (earth's as 1)	0.0769

ELEMENTS OF VENUS FOR THE 1st JAN., 1840.

Mean distance from the sun in miles	68,713,500
Same (earth's distance as 1)	.7233317
Greatest distance, same unit	.7282636
Least distance, same unit	.7183998
Eccentricity (semi-axis major as 1)	.0068183
Annual variation of same (decrease)	.0000006271
Sidereal revolution in days	224.7007754
Synodical revolution in days	583.920
Longitude of the perihelion	124° 14′ 25″.6
Annual variation of same (decrease)	3″.24
Longitude of the ascending node	75° 11′ 29″.8
Annual variation of same (decrease)	20″.50
Inclination of orbit to the ecliptic	3° 23′ 31″.4
Annual variation of same (increase)	0″.07
Mean daily motion in orbit	96′ 7″.8
Time of rotation on axis	23h 21m 21s
Inclination of axis to the ecliptic	Uncertain

Apparent diameter	17''.10
Diameter in miles	7,896
Diameter (earth's being 1)	0.925
Volume (earth's being 1)	.9960
Mass or weight (earth's being 1)	.894
Density (earth's being 1)	0.923
Light received at perihelion (earth's being 1)	1.94
Same at aphelion (earth's being 1)	1.91
Weight of a terrestrial pound, or gravity	0.90
Space fallen through in one second of time, in feet	14.5

ELEMENTS OF THE EARTH, 1st JAN., 1801.

Mean distance in miles	95,000,000
Greatest distance (mean distance 1)	1.0167751
Least distance, same unit	0.9832249
Mean sidereal revolution (solar days)	365d 06h 09m 09s.6
Mean tropical revolution	365d 05h 48m 49s.7
Mean annualistic revolution	365d 06h 13m 49s.3
Revolution of the sun's perigee (solar days)	7,645,793
Mean Longitude (20'' for aberration)	100° 39' 10''.2
Earth's motion in perihelio in a mean solar day	1° 01' 09''.9
Mean motion in a solar day	0° 59' 08''.33
Mean motion in a sidereal day	0° 59' 58''.64
Motion in aphelion in a mean solar day	0° 57' 11''.50
Mean longitude of perihelion	99° 30' 05''.0
Annual motion of perihelion (east)	11''.8
Same referred to the ecliptic	1' 01''.9
Complete tropical revolution of same in years	20,984
Obliquity of the ecliptic	23° 27' 56''.5
Annual diminution of same	0''.457
Nutation (semi-axis major)	9''.4
Precession (annual); luni-solar	50''.4
Precession in longitude	50''.1
Complete revolution of vernal equinox in years	25,868
Lunar nutation in longitude	17''.579
Solar nutation in longitude	1''.137
Eccentricity of orbit (semi-axis major 1)	0.01678356
Annual decrease	0.0000004163
Daily acceleration of sidereal over mean solar time	3' 55''.91
From vernal equinox to summer solstice	92d 21h 50m
From summer solstice to autumnal equinox	93d 13h 44m.

From autumnal equinox to winter solstice	89d 16h 44m
From winter solsticé to vernal equinox	89d 01h 33m
Mass (sun as 1)	0.0000028173
Density (water as 1)	5.6747
Mean diameter in miles	7916
Polar, " "	7898
Equatorial "	7924
Centrifugal force at equator	0.00346
Light arrives from the sun in	8′ 13″.3
Aberration	20″.25

ELEMENTS OF THE MOON.—EPOCH 1st JAN., 1801.

Mean distance from the earth (earth's radius 1)	60.273433
Mean sidereal revolution in days	27.321661
Mean synodical revolution in days	29.5305887
Eccentricity of orbit	0.054908070
Mean revolution of nodes in days	6793.391080
Mean revolution of apogee in days	3232.575343
Mean longitude of node at epoch	13° 53′ 17″.7
Mean longitude of perigee	266° 10′ 07″.5
Mean inclination of orbit	5° 08′ 39″.96
Mean longitude of moon at epoch	118° 17′ 08″.3
Mass (earth as 1)	0.011364
Diameter in miles	2164.6
Density (earth as 1)	0.556
Gravity or weight of one terrestrial pound	0.16
Bodies fall in one second, in feet	2.6
Diameter (earth as 1)	0.264
Density (water as 1)	3.37
Inclination of axis	1° 30′ 10″.8
Maximum evection	1° 20′ 29″.9
" variation	35′ 42″.0
" annual equation	11′ 12″.0
" horizontal parallax	1° 01′ 24″.0
Mean " "	57′ 00″.9
Minimum " "	53′ 48″.0
Maximum, apparent diameter	33′ 31″.1
Mean " "	31′ 07″.0
Minimum " "	29′ 21″.9

ELEMENTS OF MARS FOR THE 1st JAN., 1840.

Mean distance from the sun in miles	145,750,000
Same (earth's distance as 1)	1.523691
Greatest distance, same unit	1.6657795
Least distance, same unit	1.3816025
Eccentricity (semi-axis major as 1)	.0932528
Annual variation of same (increase)	.0000090176
Sidereal revolution in days	686.9794561
Synodical revolution in days	779.836
Longitude of the perihelion	333° 6′ 38″.4
Annual variation of same (increase)	15″.46
Longitude of the ascending node	48° 16′ 18″.0
Annual variation of same (decrease)	25″.22
Inclination of orbit to the ecliptic	1° 51′ 5″.7
Annual variation of same (decrease)	0″.01
Mean daily motion in orbit	31′ 26″.7
Time of rotation on axis	24h 37m 22s
Inclination of axis to the ecliptic	59° 41′ 49″
Apparent diameter	5″.8
Diameter in miles	4,070
Diameter (earth's being 1)	0.519
Volume (earth's being 1)	.1364
Mass or weight (earth's being 1)	0.134
Density (earth's being 1)	0.948
Light received at perihelion (earth's being 1)	.524
Same at aphelion (earth's being 1)	.360
Weight of a terrestrial pound or gravity	0.49
Space fallen through in one second of time, in feet	7.9

ELEMENTS OF THE ASTEROIDS.

Name.	Distance from the Sun. Earth's 1.	Period of revolution in days.	Eccentricity of orbit.	Inclination of orbit to ecliptic.	Longitude of ascend. node.	Longitude of perihelion.	Longitude at epoch.	Epoch. G. Greenwich. B. Berlin.
Ceres..............	2.7671120	1681.271	0.0798589	10° 36′ 30″.9	80° 48′ 57″.5	149° 25′ 51″.4	249° 31′ 26″.2	10 June, '58
Pallas	2.7708890	1684.258	0.2898672	34° 42′ 29″.8	172° 38′ 32″.7	122° 7′ 38″.4	224° 28′ 25″.5	29 May, '58
Juno...............	2.6686796	1592.368	0.2561735	13° 3′ 9″.8	170° 53′ 22″.0	54° 0′ 55″.8	104° 2′ 31″.1	29 Jan., '58
Vesta..............	2.3618405	1325.366	0.0902088	7° 8′ 9″.1	103° 21′ 10″.8	250° 35′ 29″.4	218° 26′ 1″.1	23 April, 58
Astræ..............	2.5777718	1511.696	0.1862906	5° 19′ 8″.1	141° 26′ 15″.7	135° 5′ 40″.3	159° 43′ 26″.6	1 March, '59
Hebe...............	2.4260617	1380.227	0.2013944	14° 46′ 24″.5	138° 35′ 29″.0	15° 7′ 48″.8	242° 1′ 33″.1	8 May, '58
Isis...............	2.3812400	1347.076	0.2307546	5° 27′ 56″.4	259° 44′ 39″.0	41° 28′ 12″.8	200° 41′ 2″.0	19 April, '57
Flora..............	2.2013860	1193.004	0.1567040	5° 53′ 8″.0	110° 17′ 48″.6	32° 54′ 28″.8	68° 48′ 31″.9	1 Jan., '48
Metis..............	2.3864005	1346.488	0.1238522	5° 36′ 0″.8	68° 31′ 5″.9	71° 5′ 47″.2	72° 54′ 6″.5	5 Dec., '57
Hygeia.............	3.1593600	2051.145	0.0984962	3° 46′ 50″.5	287° 17′ 3″.5	230° 42′ 54″.1	25° 49′ 27″.8	5.5 Dec., '57
Parthenope.........	2.4525900	1402.923	0.0988870	4° 36′ 57″.6	125° 4′ 26″.9	316° 10′ 57″.5	283° 57′ 32″.2	13 Oct., '57
Victoria	2.3344916	1302.756	0.2178214	8° 23′ 1″.5	235° 33′ 34″.9	301° 56′ 36″.0	312° 39′ 52″.4	27 March, '58
Egeria.............	2.5762794	1510.383	0.0873755	16° 32′ 28″.6	43° 18′ 27″.1	119° 35′ 3″.1	259° 47′ 22″.8	2 Aug. '57
Irene..............	2.5875348	1520.292	0.1680574	9° 6′ 30″.5	86° 51′ 16″.1	178° 30′ 31″.4	89° 30′ 12″.9	15 June, '57
Eunomia............	2.6430842	1569.510	0.1878466	11° 43′ 39″.8	293° 56′ 14″.8	27° 27′ 25″.5	181° 45′ 1″.9	25 Dec., '58
Psyche	2.9228660	1825.205	0.1346338	3° 4′ 7″.6	150° 32′ 48″.5	12° 40′ 34″.3	51° 35′ 24″.3	29 Jan., '58
Thetis.............	2.4726140	1420.144	0.1265394	5° 35′ 25″.8	125° 26′ 29″.6	259° 38′ 21″.6	342° 20′ 9″.0	26 Nov., '55
Melpomene..........	2.2956321	1270.430	0.2174420	10° 8′ 58″.6	150° 4′ 30″.7	15° 9′ 58″.0	167° 39′ 20″.9	7 Sept., '57
Fortuna	2.2418279	1293.272	0.1579898	1° 32′ 33″.7	211° 25′ 0″.2	30° 23′ 30″.5	149° 16′ 50″.5	6 March, '58
Massilia...........	2.4090588	1304.397	0.1436802	0° 41′ 9″.7	206° 36′ 23″.9	98° 16′ 29″.7	54° 45′ 59″.6	6 March, '58
Lutetia............	2.4343600	1387.314	0.1622043	3° 5′ 21″.1	80° 29′ 37″.6	326° 31′ 44″.8	153° 14′ 24″.5	4 Nov. '56
Calliope...........	2.9094987	1812.699	0.1019601	13° 45′ 28″.4	66° 36′ 21″.8	56° 34′ 18″.1	224° 46′ 26″.6	24 Dec., '57
Thalia	2.6280278	1556.118	0.2314366	10° 13′ 18″.1	67° 37′ 4″.9	123° 51′ 19″.1	133° 10′ 27″.3	1 Jan., '60
Themis.............	3.1483907	2035.807	0.1165263	0° 48′ 57″.7	36° 7′ 27″.9	138° 54′ 29″.1	110° 19′ 14″.0	18 Oct., '57
Phocea.............	2.4010689	1358.349	0.2525829	21° 35′ 59″.6	214° 4′ 54″.6	302° 46′ 9″.0	294° 46′ 18″.5	24 Dec., '57
Proserpine.........	2.6555649	1580.640	0.0873697	3° 35′ 39″.4	45° 54′ 19″.7	255° 3′ 43″.2	295° 26′ 25″.9	10 July, '57
Euterpe............	2.3464554	1312.858	0.1729606	1° 35′ 31″.5	93° 49′ 58″.1	87° 29′ 25″.3	126° 41′ 17″.7	3 Aug., '58
Bellona............	2.7750890	1688.550	0.1546816	9° 22′ 32″.8	144° 48′ 14″.7	122° 18′ 36″.6	159° 1′ 57″.3	10 Feb., '58
Amphitrite.........	2.5542518	1491.053	0.0726130	6° 7′ 52″.3	356° 26′ 33″.7	56° 29′ 5″.0	180° 1′ 12″.6	1 March, '54
Urania.............	2.3687922	1324.876	0.1259758	2° 5′ 57″.4	308° 12′ 59″.2	31° 15′ 29″.0	241° 5′ 32″.1	27 March, '58
Euphrosyne.........	3.1561600	2048.030	0.2160123	26° 25′ 12″.4	31° 25′ 23″.0	93° 51′ 6″.6	53° 49′ 50″.3	15 May, '57
Pomona.............	2.5830794	1516.867	0.0817246	5° 29′ 14″.0	220° 49′ 37″.6	195° 42′ 33″.5	271° 58′ 14″.9	27 June, '57
Polyhymnia.........	2.8646120	1770.912	0.3376790	1° 56′ 48″.0	9° 14′ 30″.4	340° 41′ 55″.8	23° 5′ 48″.3	1 Jan., '55

ELEMENTS OF THE ASTEROIDS.—Continued.

Name.	Distance from the Sun. Earth's 1.	Period of revolution in days.	Eccentricity of orbit.	Inclination of orbit to ecliptic.	Longitude of ascend. node.	Longitude of perihelion.	Longitude at epoch.	Epoch. G. Greenwich. B. Berlin.
Circe..............	2.6845887	1606.575	0.1119805	5° 26′ 55″.5	184° 49′ 14″.1	147° 58′ 32″.1	198° 1′ 14″.7	9.7 April, '55
Leucothea..........	2.9786900	1878.018	0.2165085	8° 15′ 17″.7	356° 24′ 37″.9	198° 17′ 0″.2	197° 49′ 30″.5	1 May, '55
Atlanta............	2.7498900	1665.596	0.2981714	18° 42′ 9″.1	830° 9′ 29″.5	42° 28′ 47″.8	36° 21′ 7″.1	1 Jan., '56
Fides..............	2.6421470	1568.671	0.1748930	8° 7′ 10″.5	8° 9′ 37″.4	66° 4′ 28″.2	42° 35′ 5″.9	1 Jan., '56
Leda...............	2.7396845	1656.340	0.1555763	6° 58′ 31″.8	296° 27′ 47″.3	100° 40′ 28″.4	112° 55′ 7″.2	1 Jan., '56
Lætitia............	2.7692870	1683.348	0.1110748	10° 20′ 50″.7	157° 19′ 30″.9	1° 58′ 57″.6	146° 44′ 19″.6	1 Jan., '56
Harmonia...........	2.2671500	1246.846	0.0460846	4° 15′ 48″.4	93° 22′ 2″.8	2° 1′ 50″.8	122° 12′ 9″.0	1 Jan , '56
Daphne.............	2.4008	1358.384	0.20248	15° 48′ 28″.0	180° 5′ 50″.8	220° 21′ 39″.8	202° 28′ 48″.5	0.5 June, '56
Isis...............	2.4338835	1886.914	0.2229198	8° 34′ 39″.6	84° 27′ 49″.7	317° 57′ 48″.4	276° 45′ 1″.9	1 July, '56
Ariadne............	2.19904	1191.108	0.15750	3° 28′ ——	264° 45′ ——	277° 11′ ——	232° 37′ ——	18.5 May, '57
Nysa...............	2.67687	1599.700	0.45889	3° 58′ ——	127° 6′ ——	118° 48′ ——	187° 29′ ——	15.5 June, '57
Eugenia............	2.69685	1617.641	0.09128	6° 35′ ——	148° 20′ ——	208° 17′ ——	252° 36′ ——	8.5 July, '57
Hestia.............	2.45689	1406.614	0.12260	2° 18′ ——	181° 32′ ——	344° 57′ ——	319° 48′ ——	16.5 Aug., '57
Aglaia.............	2.88954	1798.988	0.14019	5° 6′ ——	4° 25′ ——	306° 1′ ——	349° 55′ ——	6 Oct., '57
Doris..............	3.10687	2000.220	0.07711	6° 30′ ——	185° 14′ ——	77° 12′ ——	359° 4′ ——	1 Nov. '57
Pales..............	3.08611	1980.255	0 23766	3° 8′ ——	290° 27′ ——	32° 49′ ——	10° 29′ ——	1 Nov., '57
Virginia...........	2.67290	1596.140	0.28904	2° 52′ ——	178° 52′ ——	6° 18′ ——	18° 42′ ——	21.4 Oct., '50

ELEMENTS OF RECENTLY DISCOVERED ASTEROIDS.

Corrected Elements of Flora and Victoria.—Berlin Mean Time.

	Virginia (50) 1858, Feb. 23.0	Nemausa (51) 1858, Jan. 0 0	Europa (52) 1858, Jan. 0.0	Calypso (54) 1858, April 8.5	Alexandra (54 1858, Dec. 30.0	Pseudo-Daphne(56) 1857, Sept. 13.0	Mnemosyne(57) Wash. M. Time. 1860, Jan. 0.0	Flora (8) 1858, Jan. 1.0	Victoria (12) 1851, Jan. 0.0	Pandora (55) 1858, Dec. 30.0
Mean longitude.....	81 41 25.6	154 28 36.1	186 22 9.6	162 27 30.0	346 21 55.2	330 58 38.0	28 29 50.0	68 48 31.9	7 42 5.0	28 26 11.4
Long. of perihelion.	10 0 12.4	175 41 27.7	102 8 48.9	92 26 10.7	298 56 0.4	294 58 25.7	58 6 21.8	110 17 48.6	301 39 24.7	11 26 7.4
" of node......	178 32 18.7	175 39 8.2	129 57 37.8	144 4 18.7	330 50 17.5	194 52 31.4	200 5 58.1	5 58 8.0	285 34 41.7	10 57 29.3
Inclination.........	2 47 58.6	9 36 37.9	7 24 41 0	5 6 59.0	11 47 9.0	7 56 2.8	15 7 40.5	9 0 56.8	8 23 19.4	7 13 30.2
Angle of eccentricity	16 40 32.5	3 50 29.7	5 49 31.5	11 53 47.7	11 28 48.3	13 7 17.6	5 58 9.7	1086.331	12 38 44.1	8 10 6.8
Mean daily motion..	823″.1440	973.8489	649.8244	897.870	796.8741	854.4862	633.006		994.8841	778.8975
Log. semi-major axis	0.423021	0.874848	0.491574	0.418060	0.482598	0.412201	0.499066	0.342696	0.868171	0.440882
Computer.........	Dr. Foster.	M. Tietjen.	M. Murmann.	Mr. Linsser.	Dr. Shultz.	Dr. Luther.	Au. Sonntag.	r. Brünow.	Dr. Brünow.	Dr. Möller.

ELEMENTS OF JUPITER FOR THE 1st JAN., 1540.

Mean distance from the sun in miles	494,256,000
Same (earth's distance as 1)	5.202767
Greatest distance, same unit	5.453663
Least distance, same unit	4.951871
Eccentricity (semi-axis major as 1)	.0482235
Annual variation of same (increase)	.000001593
Sidereal revolution in days	4332.5848032
Synodical revolution in days	398.867
Longitude of the perihelion	11° 45′ 32″.8
Annual variation of same (increase)	6″.65
Longitude of the ascending node	98° 48′ 37″.8
Annual variation of same (decrease)	15″.90
Inclination of orbit to the ecliptic	1° 18′ 42″.4
Annual variation of same (decrease)	0″.23
Mean daily motion in orbit	4′ 59″.3
Time of rotation on axis	9h 55m 26s
Inclination of axis to the ecliptic	86° 54′ 30″
Apparent diameter	38″.4
Diameter in miles	92,164
Diameter (earth's being 1)	11.225
Volume (earth's being 1)	1491.
Mass or weight (earth's being 1)	342.738
Density (earth's being 1)	0.238
Light received at perihelion (earth's being 1)	.0408
Same at aphelion (earth's being 1)	.0336
Weight of a terrestrial pound or gravity	2.45
Space fallen through in one second of time in feet	39.4

ELEMENTS OF JUPITER'S SATELLITES.

No. 1. No Eccentricity.

Sidereal revolution in days	1d 18h 27m 33.506s
Mean distance (Jupiter's radius 1)	6.04853
Inclination of orbit to a fixed plane	0° 00′ 00″.0
Inclination of this plane to Jupiter's equator	0° 00′ 06″.0
Mass, that of Jupiter being 1,000,000,000	17328

No. 2. No Eccentricity.

Sidereal revolution in days	3d 13h 14m 36.393s
Mean distance (Jupiter's radius 1)	9.62347

Inclination of orbit to a fixed plane............... 0° 27′ 50″
Inclination of this plane to Jupiter's equator....... 0° 01′ 05″
Retrograde revolution of nodes on fixed plane in years 29.9142
Mass, Jupiter's being 1,000,000,000................. 25235

No. 3. Eccentricity Small.

Sidereal revolution in days.............. 7d 03h 42m 33.362s
Mean distance (Jupiter's radius 1).................. 15.35024
Inclination of orbit to a fixed plane............... 0° 12′ 20″
Inclination of this plane to Jupiter's equator....... 0° 05′ 02″
Retrograde revolution of nodes on fixed plane in years 141.7390

No. 4. Eccentricity Small.

Sidereal revolution in days............. 16d 16h 31m 49.702s
Mean distance (Jupiter's radius being 1)............ 26.99835
Inclination of orbit to a fixed plane............... 0° 14′ 58″
Inclination of this plane to Jupiter's equator....... 0° 24′ 04″
Retrograde revolution of node on fixed plane in years. 531,000

ELEMENTS OF SATURN FOR THE 1st JAN., 1840.

Mean distance from the sun in miles................... 906,205,000
Same (earth's distance as 1).......................... 9.538850
Greatest distance, same unit.......................... 10.073278
Least distance, same unit............................. 9.004422
Eccentricity (semi-axis major as 1)................... .0560265
Annual variation of same (decrease)................... 0.000003124
Sidereal revolution in days........................... 10759.2197106
Synodical revolution in days.......................... 378.090
Longitude of the perihelion........................... 89° 54′ 41″.2
Annual variation of same (increase)................... 19″.31
Longitude of the ascending node....................... 112° 16′ 34″.2
Annual variation of same (decrease)................... 19″.54
Inclination of orbit to the ecliptic.................. 2° 29′ 29″.9
Annual variation of same (decrease)................... 0″.15
Mean daily motion in orbit............................ 2′ 0″.6
Time of rotation on axis.............................. 10h 29m 17s
Inclination of axis to the ecliptic................... 61° 49′
Apparent diameter..................................... 17″.1
Diameter in miles..................................... 75,070
Diameter (earth's being 1)............................ 9.022
Volume (earth's being 1).............................. 772.0
Mass or weight (earth's being 1)...................... 102.682

Density (earth's being 1).. 0.138
Light received at perihelion (earth's being 1)................. .0123
Same at aphelion (earth's being 1).............................. .0099
Weight of a terrestrial pound or gravity........................ 1.09
Space fallen through in one second of time in feet.............. 17.6

ELEMENTS OF SATURN'S SATELLITES.

No. 1. Mimas.

Sidereal revolution in days.................. 0d 22h 37m 27.9s
Mean distance (Saturn's radius 1)..................... 3.3607
Epoch.. 1790.0
Mean longitude at epoch......................... 256° 58′ 48″
Eccentricity and Peri-Saturnium.................. Unknown.

No. 2. Enceladus.

Sidereal revolution in days.................. 1d 08h 53m 06.7s
Mean distance (Saturn's radius 1)..................... 4.3125
Epoch.. 1836.0
Mean longitude at epoch......................... 67° 41′ 36″
Eccentricity and Peri-Saturnium.................. Unknown.

No. 3. Tethys.

Sidereal revolution in days.................. 1d 21h 18m 25.7s
Mean distance (Saturn's radius 1)..................... 5.3396
Epoch.. 1836.0
Mean longitude at epoch......................... 313° 43′ 48″
Eccentricity and Peri-Saturnium.................. Uncertain.

No. 4. Dione.

Sidereal revolution in days.................. 2d 17h 41m 08.9s
Mean distance (Saturn's radius 1)..................... 6.8398
Epoch.. 1836 0
Mean longitude at epoch......................... 327° 40′ 48″
Eccentricity and Peri-Saturnium.................. Uncertain.

No. 5. Rhea.

Sidereal revolution in days.................. 4d 12h 25m 10.8s
Mean distance (Saturn's radius 1)..................... 9.5528
Epoch.. 1836.0
Longitude at epoch.............................. 353° 44′ 00″
Eccentricity and Peri-Saturnium.................. Uncertain.

No. 6. Titan.

Sidereal revolution.......................... 15d 22h 41m 25.2s
Mean distance (Saturn's radius 1)..................... 22.1450

Epoch 1830.0
Mean longitude at epoch .... 137° 21′ 24″
Eccentricity 0.02934
Longitude of Peri-Saturnium .... 256° 38′ 11″

No. 7. HYPERION.

Sidereal revolution 21d 07h 07m 40.8s
Mean distance (Saturn's radius 1) 26.7834
Other elements unknown.
Discovered (Sept. 19, 1848,) by Bond of Cambridge, and by Lassell, of Liverpool.

No. 8. JAPETUS.

Sidereal revolution 79d 7h 53m 40.4s
Mean distance (Saturn's radius 1) 64.3590
Epoch 1790.0
Mean longitude at epoch .... 269° 37′ 48″
Eccentricity and Peri-Saturnium .... Unknown.

ELEMENTS OF URANUS FOR THE 1st JAN., 1840.

Mean distance from the sun in miles 1,822,328,000
Same (earth's distance as 1) 19.18239
Greatest distance, same unit 20.07630
Least distance, same unit 18.28848
Eccentricity (semi-axis major as 1) .... .0466006
Annual variation of same .... Unknown.
Sidereal revolution in days 30686.8205556
Synodical revolution in days 369.656
Longitude of the perihelion .... 168° 5′ 24″
Annual variation of same (increase) .... 2″.28
Longitude of the ascending node .... 73° 8′ 47″.8
Annual variation of same (decrease) .... 36″.05
Inclination of orbit to the ecliptic .... 0° 46′ 29″.2
Annual variation of same (increase) .... 0″.03
Mean daily motion in orbit .... 42″.4
Time of rotation on axis .... Unknown.
Inclination of axis to the ecliptic .... Unknown.
Apparent diameter .... 4″.1
Diameter in miles 36,216
Diameter (earth's being 1) 4.344
Volume (earth's being 1) 86.5
Mass or weight (earth's being 1) 17.551

Density (earth's being 1).................................. 0.180
Light received at perihelion (earth's being 1)................. .0027
Same at aphelion (earth's being 1).......................... .0025
Weight of a terrestrial pound or gravity....................... 0.76
Space fallen through in one second of time, in feet............ 12.3

ELEMENTS OF URANUS' SATELLITES.

No. 1. Ariel.
 Sidereal revolution in days.............. 2d 12h 29m 20.66s
 Mean distance.................................... 7.40

No. 2. Umbriel.
 Sidereal revolution in days................ 4d 3h 28m 8.00s
 Mean distance 10.31

No. 3. Titania.
 Sidereal revolution in days.............. 8d 16h 56m 31.30s
 Mean distance.................................... 16.92

No. 4. Oberon.
 Sidereal revolution in days............... 13d 11h 7m 12.6s
 Mean distance.................................... 22.56

ELEMENTS OF NEPTUNE FOR THE 1st JAN., 1854.

Mean distance from the sun in miles................. 2,853,420,000
Same (earth's distance as 1)........................... 30.03627
Greatest distance, same unit........................... 30.29816
Least distance, same unit.............................. 29.77438
Eccentricity (semi-axis major as 1).................... .0087103
Annual variation of same............................. Unknown.
Sidereal revolution in days............................ 60126.722
Synodical revolution in days........................... 367.488
Longitude of the perihelion.......................... 47° 17′ 58″
Annual variation of same............................. Unknown.
Longitude of the ascending node................... 130° 10′ 12″.3
Annual variation of same............................. Unknown.
Inclination of orbit to the ecliptic................. 1° 46′ 59″.0
Annual variation of same............................. Unknown.
Mean daily motion in orbit............................. 21″.6
Time of rotation on axis............................. Unknown.
Inclination of axis to the ecliptic.................. Unknown.

Apparent diameter.. 2″.4
Diameter in miles 33,610
Diameter (earth's being 1).. 4.719
Volume (earth's being 1).. 76.6
Mass or weight (earth's being 1).. 19.145
Density (earth's being 1).. 0.222
Light received at perihelion (earth's being 1).. .0011
Same at aphelion (earth's being 1).. .0011
Weight of a terrestrial pound or gravity.. 1.36
Space fallen through in one second of time, in feet.. 21.8

ELEMENTS OF NEPTUNE'S SATELLITES.

No. 1.

Sidereal revolution.. 5d 21h 2m 43s
Longitude of the ascending node.. 175° 40′
Longitude of perihelion.. 177° 30′
Inclination to ecliptic.. 151° 0′
Eccentricity.. 0.10597

ELEMENTS OF PERIODICAL COMETS.

HALLEY'S COMET, 1835, NOV. 15.

Time of perihelion passage.. 22h 41m 22s
Longitude of perihelion.. 304° 31′ 32″
Longitude of the ascending node.. 55° 9′ 59″
Inclination to the ecliptic.. 17° 45′ 5″
The semi-axis.. 17.98796
Eccentricity.. 0.967391
Period in days.. 27,865d.74
Retrograde..

ENCKE'S COMET, 1845, AUG. 9.

Time of perihelion passage.. 15h 11m 11s
Longitude of perihelion.. 157° 44′ 21″
Longitude of the ascending node.. 334° 19′ 33″
Inclination to the ecliptic.. 13° 7′ 34″
The semi-axis.. 2.21640
Eccentricity.. 0.847436
Period in days.. 1,205d.23
Direct..

Biela's Comet, 1846, Feb. 11.

Time of perihelion passage	0h 2m 50s
Longitude of perihelion	109° 5′ 47″
Longitude of the ascending node	245° 56′ 58″
Inclination to the ecliptic	12° 34′ 14″
The semi-axis	3.50182
Eccentricity	0.755471
Period in days	2,393d.52
Direct	

Faye's Comet, 1843, Oct. 17.

Time of perihelion passage	3h 42m 16s
Longitude of perihelion	49° 34′ 19″
Longitude of the ascending node	209° 29′ 19″
Inclination to the ecliptic	11° 22′ 31″
The semi-axis	3.81179
Eccentricity	0.555962
Period in days	2,718d.26
Direct	

De Vico's Comet, 1844, Sept. 2.

Time of perihelion passage	11h 36m 53s
Longitude of perihelion	342° 31′ 15″
Longitude of the ascending node	63° 49′ 31″
Inclination to the ecliptic	2° 54′ 45″
The semi-axis	3.09946
Eccentricity	0.617256
Period in days	1,993d.09
Direct	

Brorsen's Comet, 1846, Feb. 25.

Time of perihelion passage	9h 13m 35s
Longitude of perihelion	116° 28′ 34″
Longitude of the ascending node	102° 39′ 36″
Inclination to the ecliptic	30° 55′ 7″
The semi-axis	3.15021
Eccentricity	0.793629
Period in days	2,042d 24
Direct	

* This comet which was observed *double* in 1846 was still divided at its return in 1852.

www.ingramcontent.com/pod-product-compliance
Lightning Source LLC
LaVergne TN
LVHW020118110826
845151LV00001B/200